The Identification and Assessment of Occupational Health and Safety Strategies in Europe

Volume I: The National Situations

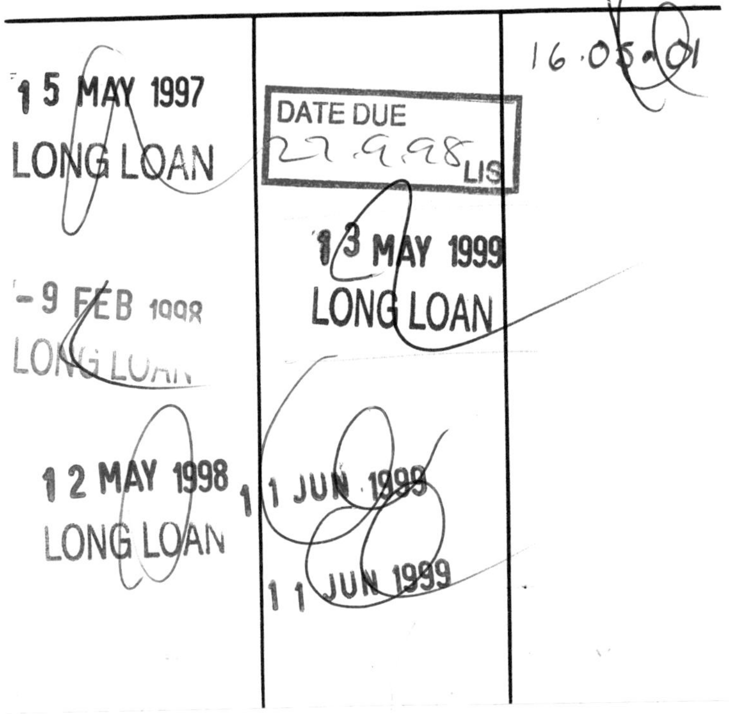

EF/96/13/EN

The Identification and Assessment of Occupational Health and Safety Strategies in Europe

Volume I: The National Situations

Edited by
David Walters
South Bank University

European Foundation
for the Improvement of Living and Working Conditions
Loughlinstown, Dublin 18, Ireland
Tel: +353 1 204 3100 Fax: +353 1 282 6456

Cataloguing data can be found at the end of this publication

Luxembourg: Office for Official Publications of the European Communities, 1996

ISBN 92-827-6641-1

Printed in Ireland

Preface

The European Work Environment Survey published by the Foundation in 1992 showed important differences among member states in working conditions related to health and safety. These inequalities persist even when economic factors are taken into account. Possible explanations of these imbalances include the health and safety strategies developed by public authorities and companies in the member states.

In 1993 the Foundation launched a project on identification and assessment of occupational health and safety strategies in Europe to produce an overview of the situation in the member states. National reports were contracted in 13 member states and two European reports were prepared. This first one describes the health and safety policies of each Member State and is based on the 13 leaflets that summarise the national reports. The second report, to be published, develops a more analytic approach. The two are complementary and will be presented as volumes I and II of the same study.

This volume is the collective effort of a team of researchers from 13 member states and has been thoroughly edited by Dr. David Walters. The complexity of the health and safety systems is high and they are in almost continuous change. Therefore inaccuracies can easily slip in the text without being noticed except by national experts. So each country chapter has been reviewed and commented by the authors of the national reports and representatives of the national governments, employers and trade unions.

We are grateful to the authors of the 13 national reports, the representatives from governments, employers and workers who reviewed them and especially to the editor.
Dr. Walters has been successful in keeping a perfect balance between the content of the original national leaflets and the need to find a standard format for all of them. His work has gone beyond the traditional editing and we deeply appreciate his efforts and enthusiasm.

We hope this volume will bring new ideas and discussions leading towards a Europe with improved health and safety.

Clive Purkiss
Director

Eric Verborgh
Deputy Director

CONTENTS

Country reports

The authors

BELGIUM:

V. De Broeck, M. De Greef, M. Heselmans, M. Van der Steen

ANPAT Institute, Brussels

DENMARK:

Poul Bitsch Olsen

Senior Researcher at the Department of Environment, Technology and Social Studies, University of Roskilde

FINLAND:

Professor Jorma Rantanen

Director General, Finnish Institute of Occupational Health, Helsinki

FRANCE:

Professor Françoise Piotet

Caisse Nationale d'Assurance Maladie (CNAM), Paris

GERMANY:

Bernhard Badura, Falko Schlottmann and Christoph Kuhlmann

Department of Social Epidemiology and Health Systems Design, School of Public Health, University of Bielefeld

GREECE:

Ilias Banoutsos, Stelios Papadopoulos, Nikos Sarafopoulos, Emmanuel Velonakis and Christina Marouli

Ergonomia Ltd, Athens

IRELAND:

Eunice McCarthy and Marie Therese Byrne

Social and Organisational Psychology Research Centre, University College Dublin

ITALY:

Dr. Graziano Frigeri

President, Società Nazionale Operatori della Prevenzione, (National Association of Prevention Professionals), Langhirano

THE NETHERLANDS:

Peter Smulders, Max van Dormolen, Erik de Gier, Michiel Kompier and Rob de Winter

Nederlands Instituut voor Praeventieve Gezondheidszorg, Leiden

PORTUGAL:

Professor Salvador Massano Cardoso

Instituto de Higiene e Medicina Social, Faculdade de Medicina de Coimbra

SPAIN:

Professor Louis Lemkow

Department of Sociology, Universidad Autónoma de Barcelona

SWEDEN:

Ingrid Stymne

Swedish Council for Work Life Research, Stockholm

UK:

David Walters

Centre for Industrial and Environmental Safety and Health, South Bank University, London

Introduction

This book is one result of a study commissioned by the European Foundation for the Improvement of Living and Working Conditions into national health and safety policies and strategies in thirteen member states of the European Union. The same study has resulted in a number of other publications, which, although each is intended to stand alone, are all very much related. The full set of publications from this study include detailed reports for each country and a second consolidated report written by Professor Françoise Piotet which takes an analytical approach to the same material that is presented from a more factual and descriptive perspective in this volume. Each consolidated report is accompanied by a leaflet that highlights some of the main issues raised in the reports.

The full national reports will be published by the European Foundation as Working Papers and the reader is referred to these for further details of issues of interest which can only be presented in outline in this volume.

Aims of the study

The goal of the study was to produce an overview and assessment of the policies and strategies for improving working conditions related to health and safety in the thirteen countries. Its aims were:

- To identify at appropriate levels policies intended to achieve healthy working conditions.
- To assess the impact of these policies on occupational accidents and diseases, absenteeism, performance, and human and social costs.
- To highlight the role of the social partners in these policies.
- To identify facilitating factors and obstacles in relation to these policies.
- To consider future trends and developments of policies and strategies intended to achieve healthy working conditions.

The Foundation's European survey on working conditions related to health and safety in 1992 suggested that northern Europe benefited from a better work environment than southern Europe. Although some differences between member states can be explained by economic and cultural factors, differences in public policies also need to be taken into account. It is therefore important to identify and assess present national strategies and policies in health and safety, and to help inform future strategy. It was for this reason that the project reported here was undertaken.

Content

The study examined the evidence of national strategies in occupational health and safety in Belgium, Denmark, Finland, France, Germany, Greece, Ireland, Italy, the Netherlands, Portugal, Spain, Sweden and the United Kingdom. Each country was studied by a national specialist who prepared the full report already mentioned and, on the basis of this, also prepared a summary document containing the main points of the detailed report. It is these summary documents that make up this volume, so that each of the chapters presents an overview of the situation in one of the thirteen countries studied, based upon the findings of the national specialist for each country and written from their perspective.

The contents of both the full reports and the summary documents contained in this book follow the same broad outline. They are divided into four main sections.

- The first section deals with the economic context of health and safety strategies and broadly considers economic structures and recent trends in the economic profile, labour market and industrial relations in each country.

- The second section is concerned with the infrastructure for occupational health and safety and looks at legislation, the regulatory agencies and inspection, provision of information, training and research. It also considers the structure and function of occupational health services within enterprises, regionally and nationally where relevant. It examines the role of occupational health insurance and economic incentives in the regulation of health and safety.
- The third section focuses on outcomes in health and safety and contains information on occupational accidents, occupational ill-health and absenteeism, as well as information on their economic costs, where it is available.

- The final section contains the assessment of the contributors to occupational health and safety strategies in their respective countries. Generally, these assessments are based upon the views of the social partners, policy makers, and health and safety professionals, as well as on the evidence presented in the previous three sections. This section tries to identify the main concerns in each country, the strengths and weaknesses of the health and safety infrastructure and policies, the priorities for action in the future and the impact of the economic situation.

The influence and consequences of membership of the European Union is reflected throughout the reports on each country.

Methods

The study was commissioned in November 1993 and carried out during the first half of 1994. It was undertaken by specialists in health and safety research from each of the countries concerned. A list of contributors and the countries for which they were responsible is found on pages ii–iv. Prior to the commissioning of the national reports, a working group consisting of the national specialists and several additional advisors as well as research managers from the European Foundation met to consider the plan of research that the Foundation had developed at that time. As a result of this meeting, a research framework was agreed which could be applied in all of the countries studied. The national reports were based upon this agreed framework. They review developments in each country during the previous five years, that is, they cover the period from 1989 to 1994, although some data will also refer to earlier periods where more recent data is not available. Their structure is summarised in the preceding paragraph.

The methods used to collect the necessary information were a mixture of documentary research and interviews with key participants.

Interview subjects were generally national representatives of trade unions and employers' organisations, representatives of health and safety professional organisations, policy makers and regulatory authorities, as well as a variety of individuals with specialist interests in health and safety matters. The interviews were based largely on the detailed brief for Section 4 of the project and were carried out to elicit information on the main concerns of the interviewees; their perceptions of the strengths and weaknesses of the occupational health and safety policies and strategies in their countries; their priorities for future actions in health and safety; their views about the role of the European Union; their assessment of the impact of the economic situation on the development of health and safety strategies in their countries.

Editorial issues

This consolidated report presents the summaries of thirteen very different contributions. Although the contributors worked to the same brief, it was not so prescriptive as to prevent considerable variation in the presentation of findings. This is partly because the very short time-scale for the production of the reports meant that the contributors were restricted in the main to secondary sources, which limited the amount of control that the study design was able to impose on the sources and collection of data. However, it is also a reflection of the enormous degree of variation in the range of countries examined in terms of their history, economy, politics and culture, all of which play a role in creating the context for national strategies on occupational health and safety and also influence their development.

Editorially, it would appear that there are two possibilities in this situation. Either an attempt can be made to fit the diverse range of material contained in the individual country reports into a rigorous, predetermined framework so that some degree of comparison can be made between material from one country and that of another, or the reports can be fitted to a much less prescriptive framework which allows the reader to identify some of the key areas that are addressed in all countries, but which at the same time retains the national diversity and idiosyncracies present in the original contributions. In this volume the latter approach has been adopted. The decision to do this was based on the understanding that in such a diverse range of cultures as that presented in the national contributions, for many well documented reasons, there is very little that is actually directly comparable in health and safety strategies, infrastructures and outcomes, without it being subject to enormous qualification. And even when the problems of differences in reporting and recording systems, and differences in the laws and regulatory practices have been accounted for, the question still remains whether it is meaningful to compare, for example, industrial accident performance in a southern European country with that of a northern European country, when their economies represent totally different stages and progression in the development of industrialisation.

On the other hand, perhaps it is useful to try to present a very broad framework for analysis of national experiences alongside one another; so, for example, it is possible to look at some of the indicators of the economic context in which health and safety strategies are developed, and to note that there are a number of characteristic features in the trends of the economies of European countries in recent years. Or it is possible to identify several legislative models of worker representation in health and safety in Europe, but pose a question about whether it is the legislative model that determines the operational effectiveness of worker representation or other features of the industrial relations context in which the operation takes place. Similarly, other examples of this typology include a range of models of occupational health services, several different models of occupational health insurance and several forms of structure and function attributed to the regulatory agencies that can be distinguished in the thirteen countries.

A further editorial issue concerns the timing of the work. There are two points that should be mentioned. The first is that the period under review, 1989–1994, has been a time of considerable legislative activity in all European Union countries. Indeed, this activity, brought about by the need to implement the Framework Directive 89/391 and its daughter Directives, is one of the characteristic developments of the period. However, member states have proceeded at different paces and in several countries implementation into national laws is still in progress. This raises the second issue of timing, which is the editorial decision on when to stop including new material in this consolidated report.

As has already been stated, the national reports were produced during the first half of 1994. The present volume was completed over a year later. Developments that were not included in the national reports but which are thought to be particularly significant, such as implementation of new framework legislation or membership of the EU, have been included in this volume. Considerable editorial discretion has been used, however, and it is hoped that a balance has been struck between retaining the national contributors' assessment of strategy over the whole of the review period and the inclusion of new developments that are pertinent to these strategies in health and safety.

Some of the issues identified in the report

Although the main purpose of this volume is to provide a descriptive and factual supplement to Professor Piotet's accompanying analysis of the development of policies and strategies in health and safety in Europe, there are several issues that emerge from the country reports included here that it is perhaps appropriate to highlight in this Introduction.

It has already been pointed out that there are broad similarities in the economic context in which recent development of health and safety strategies has taken place, with the recession of the early 1990s continuing to exert a powerful effect on the economy in all of the countries studied. The direct and indirect implications of national economic performance are referred to in all of the country reports. Increased unemployment, concerns over job security, precarious employment, part-time and temporary work, change from larger to smaller business units and changes in management style are all factors that feature in the country reports. The increased role of women in the labour market and the ageing workforce are also common features, particularly in northern European countries. All of these developments are relevant to the assessment of health and safety strategies. The decline in membership of trade unions has continued across Europe (with the possible exception of some of the Nordic countries) and there are very wide differences between the membership figures for different countries. This is not a directly comparative measure, however, since trade union membership implies different meanings in different cultures. Nevertheless, overall declining membership is clearly an issue for trade unions and one which must influence their attitudes and strategies towards health and safety.

One feature of health and safety structures that is apparent in many countries is that the development of the primary legislation that is responsible for defining regulatory boundaries is more related to previous eras of industrialisation than to the present. Another related issue that features in some of the national reports is the limits of the regulatory agencies themselves, with the assertion that such agencies are already greatly underfunded and overstreched by their regulatory duties. The impact of recent EU Directives in the process of the modernisation of the regulatory approach is referred to in many of the reports, but the impression is given that generally the legislation in many countries has not kept pace with the recent rapid changes that are occurring in the organisation of employment and production. As previously pointed out, it is possible to identify several approaches to labour inspection and to the provision of occupational health services as well as to health and safety insurance systems. While countries can be grouped according to these models, it is not clear from the national reports whether some models have the potential for adoption elsewhere, or, indeed, whether some can be regarded as more effective than others. Again, there are enormous problems of culture and context that need to be fully explored before meaningful comparisons can be made.

Training in health and safety is a feature that is clearly of national concern in many countries and there appears to be a huge range in the quality and quantity of provision and of variation between countries. It is apparent in most countries that there is limited data on provision, so although training is known to occur it is difficult to measure the scale of provision. Some features are prominent; for example, there is very little formal provision for health and safety education in schools, although developments are beginning to take place in a small number of northern European countries. Trade unions appear to play an important role in training in many countries, not only of their own representatives, but also of other workers, as well as being involved in joint training. Training at a professional level is undergoing change with some countries in the process of introducing new systems for the recognition of training; there are also developments occurring at European level.

As far as health and safety outputs are concerned, data on occupational accidents show some already well established features, such as the hazardous nature of industries like construction and engineering. Other data from some countries, which must be interpreted with caution, suggests that proportionately more young people, temporary and foreign workers are affected by accidents.

Many countries report problems with the reliability of data on injuries and ill-health. Under-reporting is the most commonly identified failing in the data. There is also an increasing recognition of the limits of much of the legally required reporting of occupational ill-health, concurrent with a growing awareness of the broadening of the definition of work-related ill-health amongst the social partners and specialists. Concern in this respect also relates to the limited ability of existing systems to reflect accurately the extent of such issues as stress and the impact of psycho-social variables upon health at work. In some of the northern European countries a strong characteristic of recent years has been the considerable interest in the economic costs of occupational accidents and diseases. It seems very likely that under the present economic conditions this interest will grow and there is certainly a need for more empirical research in this field. The cost effectiveness of preventive measures and cost benefit analysis in health and safety in general are also features that are evident in the analysis of specialists, regulatory agencies and employers in some of the northern European countries, although generally the economic aspect of health and safety performance seems less prominent overall than might have been anticipated given general economic trends.

A final point with which to end this Introduction is to return to the fact that although all of the contributors have followed the same brief, there is great variation in the degree to which their reports reflect the contents of the original brief. While this variation adds to the richness of the national reviews, it should also be clear that such variation is also a result of the limits to availability of information in different countries which in turn is often a reflection of the degree of development of health and safety infrastructures within countries. Although health and safety is clearly in a process of dynamic change in all the countries surveyed, and although the impact of the membership of the European Union is clearly an important influence on this process of change, it is also very important to acknowledge that there are very real differences in health and safety provision between countries. Harmonisation does not mean that all workers in Europe are currently experiencing the same standards of health and safety at their workplaces. This should be of very serious concern to policy makers and strategists in the development of health and safety at work in the future.

Acknowledgements

This is a consolidated report and as such contains the work of a number of different contributors. As far as the editing process is concerned, I would like to acknowledge the support and co-operation of those contributors with whom it was possible for me to discuss issues during the early stages of the drafting of this volume. I am also grateful for the support and helpful comments received from the project research manager, Mr Jaume Costa, who responded quickly and courteously to all my inquiries.

I am particularly grateful to Ms Diane West who played a significant role in the editing of this text by using her considerable experience and skills to produce a more readable version of the material included here.

Despite the debts owed to those mentioned, the final responsibility for the editing rests with me and I hope that I have managed to convey the sense of what each author intended for his contribution.

David Walters
South Bank University
July 1995

1 The context of national occupational health and safety policies in Belgium

Economic structure

At the end of June 1991, Belgium had a working population of 4,210,100, the majority of whom were employed in the private sector. Characteristic of current trends has been substantial growth in the tertiary sector, to the detriment of the secondary and primary sectors.

In June 1991 the majority of Belgian companies (97%) were small- and medium-sized enterprises (SMEs) active in manufacturing, trade, services and the professions, employing less than 50 people. This represents about 40% of all workers. SMEs, particularly those employing less than 5 people, are mainly found in the trade, catering and repair sectors.

The labour market

Unemployment

There has been a slowing down of economic activity. As a consequence, the number of jobs created has decreased and unemployment increased, which is particularly evident amongst males, young persons and in the increase of short-term unemployment. Table 1 shows the number of those unemployed distributed by sex, length of unemployment and nationality.

Total number of totally unemployed registered 1993	475,867 (100%)
Females	273,553 (42%)
Males	202,314 (57%)
Unemployed under 1 year	43.2%
Unemployed between 1 and 2 years	19.2%
Unemployed more than 2 years	37.6%
Belgian nationality	84.1%
Foreign nationality	15.9%

Table 1: Number unemployed by sex, length of unemployment and nationality

In order to reduce unemployment, the authorities have provided a series of schemes designed to get the unemployed back to work, including youth employment through apprenticeships, the 'troisième circuit' (which provides jobs in the non-commercial sector), subsidised contracts and bonuses.

Women and part-time work

In Belgium, some 465,966 people, the majority of whom are women, work part time (1992). The proportion of part-time jobs out of the total number of jobs in the country has grown from 3% in 1973 to 12% in 1992. However, the great majority of part-time jobs (90%) are in the tertiary sector where 85% of all working women are employed. Women amount to about 40% of the total workforce, but part-time work and other forms of atypical employment such as fixed-period contracts, temporary replacements and so on, serve to reinforce the insecure nature of women's work. The proportion of males in employment is at its maximum in the age-group 25–49. For females, the highest figures are in the 25–39 age-group, after which there is a systematic decline in the proportion of women in employment; 66% of women with part-time jobs are between the ages of 20 and 40.

Moonlighting

There are no official figures for moonlighting, but, according to the official statistics, over 15% of GNP is generated by this means. It has been estimated that the phenomenon could involve between 300,000 and 400,000 jobs.

Foreign workers

In 1989 Belgium had some 196,436 foreign workers, the great majority of whom were men; 72% were workers from other European countries. They mainly worked in the service (51.5%) and secondary sectors (47.8%).

Industrial relations

Collective labour agreements (CCTs) are the subject of negotiations between union representatives and the employer within the company, and between worker representative organisations and employers at other levels (sectorial and federal).

Within the company, discussions take place at three levels: in the workers council, in the safety, hygiene and the work environment (SHE) committee and through the union representatives.

- A works council must be instituted in any company employing 100 or more workers; its role is essentially consultative.
- In any company employing over 50 people an SHE committee must be set up; this is composed of representatives of both employees and management.
- Union delegates represent the workers and assume certain of the attributes of the works council and of the SHE committee if these bodies have not been instituted.
- Employers are affiliated to professional sectorial organisations which are all served by the Fédération des Entreprises de Belgique (FEB) which speaks for the employers at federal level. In addition, employers may be affiliated to the NCMV (de organisatie voor zelfstandige ondernemers) in Flanders or the UCM (Union des Classes Moyennes) in the Walloon. Both organisations represent the

interests at federal level of small and medium-sized enterprises (SMEs) and the self-employed.

Consultation at the sectorial level is undertaken by parity committees composed of an equal number of representatives of workers and employers. At federal level, consultation takes place within the National Labour Council where the social partners may sign collective labour agreements whose influence is often nationwide and intersectorial. Health and safety problems are always discussed by the Higher Council for Safety, Hygiene and the Work Environment.

2 Occupational health and safety policies and structures

Legislation

Belgian legislation on safety and health at work has gone through a long formation and development period. Over the past few years, however, this has been distilled into the following basic principles:

1. The issue of safety and health at work should be approached in a structural manner in order that detailed regulations become unnecessary and it will not be necessary to regulate everything.

2. Work situations, activities and equipment associated with serious risks must be made the subject of special regulations.

3. If the technological developments concerned are still evolving, the regulations should be aimed at the end result. This should be complemented, however, by specific recommendations which, without necessarily creating any obligation, do create a presumption of respect for the regulations.

4 As well as being subject to special regulations, it should be compulsory to notify work situations, activities and equipment associated with very serious risks to the authorities enforcing them, and that these should be the subject of prior authorisation.

Problems and workers concerned

The regulations apply to all workers, whether with or without a work contract, who carry out work under the authority or control of others. Domestic staff are outside the scope of the regulations.

Current Belgian legislation is essentially based on two fundamental laws: the Act of 10 June 1952 governing safety and health at the workplace and the Act of 11 July 1961 governing dangerous machinery and protective measures. The majority of the transcriptions of European Directives are based upon these Acts. The relevant royal decrees have been combined to form the General Regulations for the Protection of Workers (RGPT).

The provisions of the European Framework Directive 89/391/EEC - although already part of Belgian legislation - have been repeated in a legislative proposal on 'the well-being of workers in the performance of their work'.

Occupational health and safety structures

There are a number of important consultative bodies and compulsory services at various levels.

1 The Higher Council for Safety, Hygiene and the Work Environment brings the social partners together on an equal basis. The Council advises the Minister on matters of policy and especially on legislation as it is being prepared.

2 Sectorial committees on safety and hygiene which formulate comments and recommendations relating to legislation are found in the construction, metal and chemical industries.

3 At company level, there are SHE committees which are compulsory in workplaces with 50 or more employees. Employees' representatives are elected every four years. The employer's representatives are appointed by the employer from management personnel. The occupational health physicion and safety officer attend in a consultative capacity. The powers of the SHE committee are laid down by the General Regulations for the Protection of Workers (RGPT) and mainly relate to rights of the SHE committee to information, advice, involvement and prior agreement.

4 Every employer must have a Safety, Hygiene and Work Environment Service (SHE service). For companies with less than 20 employees, the employer himself can be the head of the SHE service.

5 The employer is obliged to call upon the official organisations to oversee certain well-defined technical operations. These operations relate to electrical installations, the lifting and handling of gear, steam engines, gas containers, personal protection equipment, etc.

6 Every employer must also provide an occupational health service. If more than 50 people are employed, the employer can institute his own company medical service; if less than 50, the employer must be a member of an inter-company medical service.

7 Within the public authorities, the principle executive agencies responsible for health and safety are the Administration of the Safety at Work Service and the Administration of the Industrial Hygiene and Medical Service. Both administrations

are part of the Ministry of Employment and Labour. They draw up proposals for legislation and each has its own inspectorate. The co-ordination of the promotion of safety and health and the raising of awareness amongst workers is the responsibility of the Commissariat Général à la Promotion du Travail (Work Promotion Board - a department of the Ministry of Employment and Labour). Participation by the social partners is directed by nine provincial committees for the promotion of work and by the 'action and propaganda' section of the Higher Council. The actual work is carried out by a series of private institutions and associations.

Control and inspection

The terms of reference of the inspection services are to contribute towards the reduction of accidents and health problems in industry and the public service through the enforcment of the official regulations. Another task of the inspection services is to report any gaps in the regulations and to help the appropriate authorities to close such loopholes.

There are two arms to the inspection services:

- the Technical Inspectorate reports to the General Directorate of Safety at Work and examines working conditions in companies presenting accident risks
- the Medical Inspectorate reports to the Directorate General of Hygiene and Medicine at Work and is concerned with working conditions presenting risks to the health of workers.

The powers of inspectors are of three kinds: the power of investigation, the power of regulation and the power of prosecution. These are chiefly defined in the Act governing labour inspection of 16 November 1972 and the corresponding implementation order of 23 October 1978. Inspectors have free access to the premises which are subject to their control. The employer is obliged to co-operate during the course of any enquiry. The power of regulation tends to be implemented through verbal or written notification.

Inspectors can prescribe appropriate measures to remedy any discrepancies or forms of nuisance which they feel constitute a danger to the safety or health of workers. They can also, if circumstances demand it, take legal action. In the case of serious infringements the problem is referred for settlement to the courts (Labour Court). The summons is sent to the Labour Court judge who plays the part of a director of public prosecutions and has the power to sue or dismiss the case.

In 1992, 283 summons were issued, 225 of which were through the Technical Inspectorate. It is estimated, on the basis of previous years, that approximately 75% of these summons (or about 212) will give rise to some sanction or other. In addition, the Technical Inspectorate imposed sanctions on 93 work situations by halting the operation. The total number of sanctions for 1993 can therefore be estimated at 305, while about 100,000 companies and institutions were visited, which represents approximately 1 sanction per 300 companies visited.

Occupational health and safety services

Under the regulations, 100% of workers under contract and a large number of workers without contracts have access to the safety service (SHE service) and health service (company medical service). In practice, the operation of these two kinds of service takes place almost entirely in companies that employ more than 20 workers.

The aim of the SHE services is preventive. A distinction is made between primary, secondary and tertiary prevention. Primary prevention concentrates on the workplace, with its main objective being to adapt working conditions before accidents or health problems can occur. Secondary prevention is aimed at tracking down dangers to health and accident warning signs at an early stage. Tertiary prevention is concerned with measures designed to prevent the further development of any disorders noted and to avoid the occurrence of other problems.

The occupational health service consists of at least one occupational health physicion and one nurse or social worker. The staff of the SHE services must have adequate technical and legal qualifications. To become an occupational health physicion in future will require four years of study after the award of a medical degree, including three years of practical experience. At present, only one postgraduate year is required.

Training for safety officers is fairly precisely defined according to whether the enterprise entails serious risks (Level 1) or moderate risks (Level 2). In addition, training at Level 2 is required for assistants to the chief safety officer where the enterprise entails serious risks. Serious and moderate risks are defined on the basis of the nature of the activity (NACE Code) and the number of workers. Companies, institutions and the public service each take full responsibility for financing the safety and health services which they have set up. Companies affiliated to an inter-company medical service pay pro rata for the number of hours of medical attendance which they are obliged to provide.

There is no connection between occupational health services and public health provision. Recently, however, some occupational health services have taken initiatives in the larger area of preventive medicine and they actively contribute to promoting a healthier lifestyle (exercise, smoking, diet, stress).

Information and assistance

Legislation stipulates that the company management must take measures to ensure that workers receive all the required information laid down by the Framework Directive 89/391/EEC. In this context workers include the employees of subcontractors, temporary replacement workers, visitors, those on work experience courses and apprentices.

Management, assisted by specialists such as the safety officer and occupational health physicion, must ensure that the SHE Committee or the union representative is kept informed and consulted on a number of subjects. The SHE service and the occupational health service are obliged to submit an annual report to the Inspection Service.

Training

Legislation lays down, on the one hand, that information provided to workers should be in the form of practical instructions and, on the other, that the employer should set up a programme of special training adapted to the work station or job. This training should take place in a number of circumstances:

- as soon as an employee starts a job
- if there is any change of work station or job
- on the introduction of new equipment or any change of equipment
- on the application of any new technology
- if there is any change in the risks entailed.

The members of the SHE Committee also have the right to an appropriate course of training. This is given during working time or in accordance with the rules fixed by the CTTs (collective labour agreements). In practice, they are often organised by the unions. The employer is responsible for training prevention experts - who may also be the safety officers - who help him in co-ordinating prevention policies.

Economic incentives

The sanctions laid down by the basic legislation in 1952 vary between 130 ECU and 2,500 ECU for infringements of the regulations in respect to safety. The sanction may theoretically amount to as much as 500,000 ECU for an employer who has not organised at least a SHE service or committee. In the case of failure to do so within one year from the previous sentence, the penalty is doubled. Proposed legislation provides for sanctions from 500 to 50,000 ECU for infringements of the safety regulations.

Research

There are a number of research initiatives, including those of the inter-company medical services, the universities and so on. What is missing is a research policy and an organisation with which all the private initiatives could register. This would help to guarantee research which is better oriented to the needs of safety and health, gives better access to results and a better translation of results into preventive policy.

Occupational health insurance

As occupational accident insurance is compulsory for all employers, the system constitutes an important element of social security. It is covered by private insurance companies who are registered and controlled by the Ministry of Social Affairs. There has been a continued increase in the amount of damages paid by insurers, as shown in Table 2.

	1989	1990	1991	1992	1993
Workplace - waged	517	549	538	556	523
Workplace - salaried	70	74	77	81	88
On way to work	122	117	137	130	140
Domestic staff	1.3	1.0	1.2	0.8	1.3
Total (+ extension Act)	717	750	759	774	758

Table 2: Trends in damages paid by insurers (1989-1993) (Amounts in millions of ECU)

The majority of insurance bodies have created a prevention service whose task is to encourage the prevention of accidents at work in its subscribing companies. The work of these prevention services includes the study and analysis of occupational accidents, the study and evaluation of risks and the provision of information, training and promotion activities aimed at both workers and employers. In addition to the expenses for the operation of their own prevention services, the insurers contribute approximately 2.4 million ECU annually for prevention at federal level.

The insurance companies are under the control of the Industrial Accidents Fund (FAT). This Fund is a public institution under the aegis of the Ministry of Social Affairs and managed by a management committee composed of representatives from organisations of both employers and workers. For some years now, the FAT has also been concerned with accident prevention.

Victims of occupational diseases are covered under the Occupational Diseases Fund (FMP). This state institution is managed by a committee on which the social partners are represented. The FMP is part of social security, which means that employers are obliged to pay a certain percentage of total salaries towards its financing and operation. Although the FMP is primarily responsible for compensating victims, it also assumes several tasks directed towards prevention.

Insurance companies apply a scale of premiums. They are partly increased in relation to the indemnities paid to victims of accidents at work in the company during the preceding years and are thus related to the frequency and seriousness of accidents at the company.

3 Health and safety outputs

Occupational accidents

Occupational accidents are very carefully recorded in Belgium. The quantity of data available is therefore sufficient, even if it is not always very recent.

Any accident which happens to a worker during or through the execution of his work contract and which occasions injury is considered an occupational accident. All occupational accidents resulting in a minimum of one day's incapacity for work are included in the statistics. A distinction is then made between accidents causing temporary incapacity, permanent incapacity and death.

In order to measure trends in occupational accidents, relative figures have been used. With regard to the number of accidents, these are calculated per million hours of exposure to the risks; this figure is called the frequency rate. In order to measure the seriousness of accidents, the reference taken is the gravity rate. This relates to the number of calendar days lost per thousand hours of exposure (actual gravity). In order to keep a record of fatal accidents and accidents causing permanent incapacity, an overall number of fixed days are counted in addition; this is then called the global gravity rate.

National figures show that, after several years in which there was a trend for the rate of occupational accidents to rise, 1991 saw a turning-point and the figures for 1992 appear to confirm the trend for the rate of occupational accidents to fall. The gravity rate (both actual and global) also fell. This means not only that accidents are less numerous, but also that they are becoming less serious. This general tendency is not, however, confirmed in every sector; the frequency rate in certain sectors is much higher than average. Table 3 shows the sectors with the highest frequency rates for 1992 and the frequency rates for the same sectors from 1989 onwards.

SECTOR	1989	1990	1991	1992
Shipbuilding	113.14	133.94	160.23	259.77
Solid fuel mining and processing	-	305.50	243.65	234.60
General construction and civil enineering	120.92	131.66	154.82	139.95
Activities connected with shipping	163.23	162.93	156.23	138.34
Building construction	143.47	141.24	141.38	137.86

Table 3: Accident frequency rates in the top five sectors

If we consider the seriousness of the accidents, we can see that it tends to affect more or less the same sectors. Table 4 shows the top five global gravity rates for 1992 and the gravity rates for the same sectors from 1989 onwards.

SECTOR	1989	1990	1991	1992
Shipbuilding	4.30	3.86	4.73	10.74
Laying out and completing	5.25	4.76	4.25	7.83
Building construction	7.77	8.29	7.52	7.78
General construction and civil engineering	7.69	6.64	9.52	7.63
Public works (roads, bridges, etc.)	4.30	2.71	3.76	3.47

Table 4: Global gravity rates in the top five sectors

Larger workplaces - those with more than 200 workers - present the most favourable accident figures and this is confirmed from year to year. However, not all sectors reflect the same situation. The service sector, for example, is the best represented in this group and it clearly presents figures that are lower than average, thus influencing the group's overall figures for accidents.

Occupational diseases

There is a list of occupational diseases, established by legislation, which qualify for compensation. A disease which does not appear on the list may also qualify for compensation if it results in a direct and definite manner from the performance of occupational activity. It is in that case up to the victim or his heirs to prove the link of causality between the disease and the exposure to occupational hazards. The significant majority of compensation claims are for silicosis, with more than twice as many claims in this category as all the other nine categories of the top ten taken together. Overall compensation claims for the ten main occupational diseases have increased from 14,228 in 1988 to 15,377 in 1992 (in thousands of ECU).

Absenteeism

Researchers, doctors and unions estimate between 6% and 7% absenteeism amongst Belgian workers. Every day, therefore, there will be some 210,000 waged and salaried workers unfit for work because of illness.

Working environment surveys

At the present time, there is no method in Belgium by which to assess the quality of working conditions in companies, although one is being developed that should allow the construction of a representative table of working conditions in companies and organisations. A representative sample of businesses and institutions will be selected from each of the NACE Code sectors. These organisations will be observed by experts using a standardised set of topics covering fifty subjects. It is then planned to analyse the responses received in order to obtain a 'working conditions barometer' which will be published every year. After evaluation and readjustment, the first barometer figures are anticipated in 1996.

The economic costs of occupational accidents and diseases and absenteeism

Up to the present time, the cost of industrial accidents, occupational diseases and absenteeism in Belgium has never been calculated in any systematic fashion, although one evaluation sets the annual cost of long-term absenteeism - over 14 days for wage earners, over 30 days for salaried workers - at between 2.4 and 3 billion ECU.

4 Assessment of occupational health and safety policies

For the public service, as for employers, union organisations and prevention experts, the protection of health and safety at work is one of the priorities in industry and should be a criterion for good management. Some of the major preoccupations of these participants are:

- the changes taking place in the nature of economic activities and the consequent changes in the risks involved
- the protection of certain of the most vulnerable groups of workers (women, pregnant women, young people, student workers, temporary workers, etc.)
- awareness of the growing number of small businesses.

One of the strong points of the Belgian system is its consultative structure which is well developed, particularly in the SHE committees. Another strong point is the ability to call upon specialists. Belgium has opted for the presence of experts within industry, each company being obliged to make use of a SHE service and the services of a company doctor; there is, nevertheless, a need to address certain problems. These include:

- the lack of integration of safety and health, for example in educational courses
- the scattered initiatives in the matter of OHS structures
- financial resources for research, training and the provision of information.

Belgium has a detailed set of regulations in the area of safety and health. This is perceived as a strong point by the unions, but as a weak one by the employers' organisations who would prefer framework legislation which allows the company to take responsibility for its detailed implementation. Safety officers feel that the General Regulations for the Protection of Workers (RGPT) are too inaccessible and too detailed. They also report problems due to the (very) part-time nature of the company medical service, and the infrequency of work inspections. In addition, they are dissatisfied with the absence of compulsory training for safety officers in small businesses.

One of the priorities for the Government is to formulate new framework legislation which would deal with new areas (such as stress), open up new fields of application (such as providing health and safety cover for the self-employed), define new responsibilities and establish new prevention structures (favouring the presence of experts in small businesses). A restructuring of the regulations in a new code should follow from this, along with the redrafting of the regulations with a view to making them more user-friendly.

According to the trade unions, action should be taken on two levels: firstly on the policy level in the form of a re-evaluation of prevention policy, and secondly a direct approach to urgent and priority problems such as the high-risk sectors, the risks of occupational cancers, back ailments, the operation of prevention services in small businesses and in the administrative and public service sectors. Employers are also concerned about small businesses and in addition are seeking initiatives at sectorial level. All participants appear to be in agreement that these actions and priorities should be processed by the departments of the Ministry of Employment and Labour in consultation with the representatives of both workers and employers.

The authorities fear that the influence of the economic situation will delay reforms that have cost implications. It is the view of experts involved in prevention that the economic crisis with its high rate of unemployment has had an adverse effect on working conditions in industry. In order for companies to stay competitive, prevention policy may be prejudiced by considerations of immediate profitability. According to the employers, on the other hand, the economic factor encourages the development of a prevention policy. When adverse economic conditions mean reduced resources, it is necessary to fall back on the most judicious use of the resources available. Wherever the risks are greatest, the advantage of an improved prevention policy will also be greatest.

Trade unions argue that a safety and health policy which could be implemented on a permanent basis should be founded on the principle of risk prevention. The unions feel that the economic situation should not change these priorities. If, in spite of everything, the conflict of interest between the economic situation and the formulation of prevention policy cannot be removed, the objective of achieving good health should be accorded its full weight when evaluating diverging interests.

Government favours a European policy on safety and health, especially in regard to the setting and publication of standards. European Directives are, after all, the basis for the restructuring of the regulations and of the new framework legislation now in preparation in Belgium. The unions would also like to see far-reaching European regulations. As far as the employers' organisation is concerned there is no doubt as to the necessity of developing a European policy, but so far EU directives have mainly led to over-regulation and additional provisions in the current legislation. In their view the aim should be to make the existing measures more effective instead of introducing new rules.

The safety officers, however, are worried that harmonisation will result in a levelling down of standards causing a devaluation of prevention. They feel, nevertheless, that it would be beneficial to have a wider dissemination of knowledge with regard to laws and regulations which apply in Europe.

1 The context of national occupational health and safety policies in Denmark

Economic structure

Just over 2.5 million people are employed in Denmark. The majority, 1.7 million, are found in the service sector with industry accounting for nearly 0.68 million workers and agriculture just over 0.12 million workers. In 1993, 1.7 million worked in the private sector and 0.8 million worked in the public sector.

The labour market

Unemployment

Between 1989 and 1994 the Danish labour market was subject to widespread unemployment which was greatest amongst women, unskilled workers and migrants; it rose from 8.7% of the total workforce in 1988 to 12.4% in 1993.

Moonlighting and part-time work

The latest available figures suggest that the percentage of the total workforce engaged in part-time work fell from 17.1% in 1988 to 16.2% in 1992. According to a study published in 1992, more than 14% of the population were involved in moonlighting in 1991 at an average rate of 63 minutes per day. The construction industry in particular has been affected by fluctuating levels of activity and a high level of moonlighting, with service companies and other small companies in the construction sector being involved in a substantial amount of moonlighting. Latterly this has been reduced by the provision of local authority grants for work in the construction sector and elsewhere. As in other countries, moonlighting has become a fact of life, despite attempts in political debate to convert such activities into part of the legitimate economy.

Migrants and women workers

The working environment of migrants tends to be more impoverished than that of other groups, especially those businesses which employ large numbers of migrants with a poor grasp of the Danish language. Women tend to be focused in caring and service-orientated jobs. They are rarely to be found in top management, but in middle management their numbers are growing.

Absenteeism

Absenteeism through illness is extremely low in Denmark. Men in industry are off sick for 4% of the time, while the figure for women is 6% (including personal illness and accidents). For white-collar workers, absenteeism through illness is less than half this amount. Danes are extremely highly motivated to go to work. A living standards study of November 1990 showed that 37% of unskilled workers and 27% of skilled workers went to work every day, despite illness. The figure for top management was 40%, whilst it was 30% for white-collar workers on lower grades and in middle management.

Industrial relations

The Danish conflict-solving system is highly institutionalised and has remained relatively unchanged for many years. Trade union membership has increased steadily during recent decades; this trend was maintained during the study period. In 1988 union membership stood at 78% of the total workforce and it had increased to 81% by 1992. The membership of employers' organisations also increased during the same period.

The network of legislation governing the resolution of conflicts which has been in place for many years has not been significantly modified by membership of the EU, although a degree of uncertainty has been introduced at the prospect of the European Court using its power to decide matters which formerly would have been reached by agreement between the parties or with the Government within Denmark.

The fundamental principle of the Danish system is that production must be protected against loss and neglect. The consequence of this is that striking for the most part is illegal, except in certain outstanding circumstances. Loss of working days is therefore negligible, and has remained so even during years when the degree of conflict has been high. However, informal (illegal) conflicts that are not reported by employers are fairly common.

'Protection of production' as a labour market philosophy means that the labour market in Denmark is unusually unprotective of workers. Hourly-paid employees have no defence against dismissal and collective agreements provide for strictly limited compensation in such circumstances. Welfare and health are a long way down the agenda of these agreements. It is for this reason that the State social security system has developed, with taxation providing for social and maternity costs.

2 Occupational health and safety policies and structures

Legislation

Along with the EU's Framework Directive (89/391/EEC), the Working Environment Act enhances Danish employee protection and covers all forms of work including moonlighting, but does not apply to the self-employed. The right to refuse to carry out work that is hazardous to health will be extended in future. This will have an effect on role distribution in the safety organisations and the relationship of representatives with employees. Compulsory assessment of workplaces will place new, comprehensive responsibilities upon companies in terms of environmental protection activities. The effect of the Framework Directive may mean that some aspects of the Danish regulations, in order to ensure proper implementation, will be more strictly interpreted, either by the Ministry or by individual cases being brought before the European Court.

Occupational health and safety structures

Culture and tradition

The tripartite system is undergoing change. Until 1982 it played an active part in the working environment, but with the election of a non-socialist government, politicians began to take on a more passive role which brought the labour market organisations together. Today the system is still composed of three units, but not in the same way as previously. The lead is taken by the State administration and the Minister, followed by the social partners with their joint initiatives, and thirdly by the EU Commission and the European Court which have taken on some of the functions for which these national parties were formerly responsible.

Integration and representation

Denmark is a highly integrated society. In terms of the working environment, the consequence of this is that information is equally well disseminated to all businesses and attitudes vary only slightly. Characteristic of the Danish structure is representation. This is so all-pervading that, it is estimated, 90% of the businesses under obligation to have health and safety representation (those with 10 or more employees) and a health and safety organisations (those with 20 employees) actually do so.

The brief of the workplace health and safety system is to provide advice and initiatives. The health and safety committees are the central part of the system and are responsible for the co-ordinated management of health and safety work and advising management on matters concerning the working environment. The safety groups – the local branches of the health and safety committees – concern themselves with recording information, dealing with accidents and with prevention in practical terms.

The many safety representatives and safety groups have set their stamp on companies' working environment initiatives, but not in a uniformly positive manner. They are not a part of other co-operative systems, and for a number of reasons they contribute towards maintaining the hierarchical structure which is now considered old-fashioned in a company that regards itself as flexible. Another feature is their separateness from the workers, who tend to take their problems to the representative or supervisor in the safety organisation, rather than trying to improve matters for themselves. The positive side of safety organisations in companies is that they exist as a coherent whole and do so on a formalised basis, so that, for example, safety representatives are in a position to carry out functions such as working and consulting with the management on the basis of the legislation alone. The safety organisation is a highly reliable channel of communication.

The problem is that there is no incentive to develop dialogue because the safety organisation system is a formal one and has many ramifications so it tends to inhibit anything in the way of spontaneous initiatives aimed at making work more pleasurable. The parties are at present discussing ways in which this can be remedied. At committee level it is being emphasised that there should be a more definite division of labour between the work of the co-ordinative health and safety committees and that of the local safety groups; planning and practice should be separated in order to stimulate preventive initiatives. Presumably this will lead to an even more pronounced professionalisation of safety work and the emphasis on action rather than discussion will be reinforced.

The Labour Inspectorate (Danish Working Environment Service)

For many years the Labour Inspectorate has tried to induce companies to take preventive measures rather than merely to react to problems, accidents and absenteeism. Its approach is based on a strategic planning model to the effect that companies should have:

- an objective
- an ongoing documentation process
- cause and effect analysis
- prioritisation and the development of problem-solving approaches.

This model has expressed itself in terms of campaigns with the objective of throwing light on specific problems, and in attempts to introduce systematic control into safety organisation activities. This has not succeeded: it has neither resulted in a basic formulation of objectives by companies, nor is there an active awareness which matches, for example, strategic management in more business-orientated contexts. However, with the requirement of the EU Framework Directive that companies with hazardous working environments must have a workplace assessment, a new, written means of regulation has come into being. The Framework Directive is being implemented by the Labour Inspectorate in such a way that *all* companies requiring affiliation to the Occupational Health Services (OHS) system must have a written workplace assessment. At the same time, the Labour Inspectorate continues to insist on co-operation between companies and OHSs, requiring the OHS and the company to enter into an agreement governing their future initiatives. Unlike the workplace assessment, this is to be submitted to the Labour Inspectorate. By this means the Labour Inspectorate is exercising its right to give OHSs a more permanent role in relation to companies, making it possible for them to be integrated with workplace assessment - the new organisational instrument for the safety organisation.

The role distribution between the Labour Inspectorate and the OHS system has become clearer in that the latter - which does the information work - now enjoys greater resources, while the Labour Inspectorate, with its more modest funding, takes on a more supervisory and authoritative function. The Labour Inspectorate will continue to play its consultative part in relation to government and organisations. This has been exemplified most recently by its debate agenda concerning aims and funding for the Working Environment Year 2005.

Control and inspection

About 60,000 inspections are carried out per year, with some 20,000 injunctions being issued. Historically the Labour Inspectorate is remarkable for its capacity to ensure that regulations are observed. The most important principle formulation takes place on a local basis, through declarations of principle in actual injunctions; the development of regulations therefore proceeds irrespective of the national political complexion. While the supervision by labour inspectorates in other countries can usually be described in terms of a routine following of precedents, the authorities in Denmark also issue a series of far-reaching injunctions which enhance understanding in new areas of regulation, although supervisory control is still the most important factor.

There are deficiencies in the inspection pattern. It is clear, for example, that the Labour Inspectorate must introduce a procedure for improving preventive measures in relation to physical stresses. This alone would justify the new OHS strategy, since ergonomic stresses are usually inexpensive and possible to prevent, and are of very great significance as regards sick leave. We can therefore expect OHSs to concern themselves with this aspect of the working environment. However, it is more difficult to issue injunctions in relation to ergonomic stresses on the grounds of the regulations, so inspectors have to spend more time on these than on other more run-of-the-mill injunctions.

The number of fatal accidents has fallen, as have cases of industrial disease caused by organic solvents. Unfortunately the results of regulation cannot be seen from the accident statistics in isolation, but there is no doubt that the monitoring work is of great importance for the development of attitudes about the working environment in Denmark.

Occupational health services

The Working Environment Act obliges employers with particularly hazardous jobs to participate in an occupational health/working environment service (OHS). The OHS system has been supplied with the resources to expand its working environment activities. It is a local and private system, the structure of which is determined by ministerial order. There are essentially three types of OHS: individual company services, industry-wide services and local centres.

More and more industries are becoming affiliated, so that manual work in the manufacturing sector is now for the most part covered. Manual work in the public sector - which covers the social and health services - is also included. Much remains to be done, however, in that many service functions and administrative jobs have no OHS coverage. It is estimated that at present companies with a total workforce of around 850,000 are members of OHSs, but it is intended that at least 80% of the workforce should be covered by the year 2000.

The OHS works in tandem with the safety organisation within the company and is thus the external consultant with access to all company levels. Nevertheless, the comment frequently expressed by the OHSs is that they lack contact with the great majority of employees. They also have problems regarding the acceptance by the safety organisations of their attempts to maintain a modest degree of presence amongst rank-and-file workers.

At present there is a tendency for the OHS either to seek closer co-operation with the organisations or to choose the opposite course to try to reinforce their position in the market for working environment services. Market relations are boosted by tailoring the product, which may range from a highly specialised working environment assistance programme to a low subscription for the obligatory membership which more or less guarantees that there will be no fieldwork. Others offer a special service such as the capacity to be of service to nationwide companies active in the working environment field. The most important service in the future will be maintaining a presence and participation in the internal life of companies. This will mean a more participatory style on the part of the professionals, plus an ability to supply highly specialised information or planning when needed.

It is encouraging that there is considerable development in the capacity of the OHS to participate in the systematic and creative development of companies. There has been a general shift from a consultative function to a participatory function in the work of the companies, and OHSs have made contributions at the highest level.

OHSs have learnt to make formal contracts with companies and to make better use of their professional expertise as well as increasing their specialisation and relevant knowledge. They are also beginning to learn to distinguish their role from that of the overall responsibility of the company for occupational safety and health, to recognise the independence of departments within organisations, to work with everyone affected by the working environment and to participate appropriately while at the same time leaving decisions about solutions and their prioritisation to the client organisations.

The OHS is a major career arena in the working environment field. The staffing is multidisciplinary, including occupational therapists, physiotherapists, doctors, nurses and engineers. There is a discernable trend away from the technically centred services originally established, towards units employing a greater number of professionals with psychological, social and organisational skills.

Occupational health insurance

Employers have compulsory insurance for their health and safety reponsibilities towards their employees and society. The insurance system is mainly private; 10 to 15 insurance companies cover the private sector while the public sector is covered by its own insurance scheme. In recent years there has been much discussion about changing the insurance system into a more actively preventive one. Although examples from the way in which the insurance companies responded to the problem of organic solvents and brain damage showed that the insurance system can respond quickly with a preventive strategy to known risks, they also revealed its limitations in this area. The cost of health and safety damage is estimated to be close to 4 billion ECUs. The insurance system covers only 2.8% of this expense, therefore the motivation for prevention is not related to the economic size of the projects undertaken.

Training

Safety organisation members receive training, with all members undergoing a compulsory course of just under one week's duration. Trade unions undertake further training of safety representatives. Professional associations are also important training providers.

Scientific training is dealt with by the universities. Since research and training go hand in hand, the tendency to locate research activities at industrial hygiene research institutes has probably inhibited the development of training at university level. With growing interest in organisational processes in working environment activities and cleaner technology within the working environment, there is also an increasing enthusiasm for a genuinely cross-discipline approach to science, in terms both of training and research.

Research

Specialist funding for working environment research has been channelled through industrial research institutes. The National Institute of Occupational Health employs over 100 researchers and substantial research activity is based around occupational health clinics. In addition, a number of universities undertake a considerable amount of working environment-related research. To some extent this is orientated towards teaching and is therefore stable, but it also embraces and forms part of the debate which takes its cue from current interest in particular fields.

The policy debate about working environment research is relatively clear in relation to subject and discipline-orientated research, but tends to be limited to these areas. There is a lack of clarity, however, with regard to:

- studying and understanding the significance of the cultural context that embraces a range of special factors affecting the regulation of the labour market
- arriving at a goal-directed symbiosis with international research necessitated by the small national research budget
- the special Danish contribution to the overall development of knowledge in this field.

3 Health and safety outputs

Occupational accidents

There has been a reduction in the numbers of recorded injuries and fatalities from occupational accidents during the period of the survey as is shown in Table 5.

	Fatalities	**Total injuries**
1986	84	61,975
1989	82	45,299
1993	61	44,247

Table 5: Occupational accidents 1986–1993

Occupational diseases

A reduction in the number of cases of occupational diseases recorded by the Labour Inspectorate occurred between 1986 and 1989. This is shown in Table 6.

Year	Total recorded recognised occupational diseases
1986	16,520
1989	14,870
1993	15,655

Table 6: Occupational diseases 1986–1993

The decline in both occupational injury and diseases may illustrate some of the effects of Labour Inspectorate initiatives. It has been a particular objective of the Inspectorate to reduce occupational injuries and disease, and the impression is that its intensive control strategy has been successful. However, it has also been pointed out that reduction in the numbers of recorded injuries and diseases are not necessarily reliable indicators of improvement in the overall well-being of employees.

Work environment surveys

The Labour Inspectorate publishes information about the work environment and related risk factors with its statistics on occupational accidents and disease. Such information has been criticised for not paying attention to the social and psychological aspects of risk.

The economic cost of occupational accidents and diseases and absenteeism

All new health and safety regulations are evaluated according to an established formula which considers their economic impact as well as other benefits. Expenses related to working environment accidents and illnesses increased by 61% between 1980 and 1990, according to one report published by the public employers' organisation in 1991.

4 Assessment of occupational health and safety policies

Danish occupational health and safety strategies can be divided into:

- enforcement strategies
- compliance and informative strategies
- market and economic strategies.

The enforcement strategy has set long-term and short-term goals. The basic activity has been inspection and control by the Labour Inspectorate under the influence of the Minister of Labour and in discussion with the organisations in the labour market. The Labour Inspectorate has been the central agency for all important decisions and initiatives; however, it has relatively limited resources which have been reduced during the period in question. Studies at company level show that management and companies respond to expensive health and safety projects only when the Labour Inspectorate is involved. This observation may be related to the fact that the Danish Labour Inspectorate is very much characterised by enforcement and regulatory instruments. Hence the Danish system can be characterised as enforcement-based.

Informative strategies have been supported by resources from the industrial health service during the last decade. Whilst the effects of these strategies are difficult to evaluate, they do not cause conflicts or meet with resistance.

Market or economic strategies have not been developed very far in Denmark. Fines and sanctions are small, compensation by the insurance system is minimal and no serious funding that could be considered an economic incentive has been made available. The cost of the insurance system is largely paid by tax revenues rather than borne by companies themselves.

Values and philosophy in relation to health and safety

As a result of the Working Environment Act, improvement of the working environment has become relatively more important to Danish wage-earners in relation to themes such as pay and working hours. The working environment and pollution take the lead in uniting members' interests in the trade union movement, and play a part in maintaining the steadily rising involvement of organised labour. This is problematic because pay and the resolution of conflicts with employers is the *raison d'être* of organised labour. Since it has been realised that trade union members regard involvement and dialogue about welfare at work as very important, both employers' and employees' organisations are looking for ways of integrating the working environment and the external environment into their activities.

There are a large number of small and medium-sized businesses in Denmark. Traditionally, the greatest concern for the working environment has been with regard to medium-sized private enterprises, which have been the object of monitoring and organisational measures. The development of 'moral' attitudes to the working environment as the result of the focus on this rather heterogeneous group of companies has been particularly noteworthy.

Supervision and enforcement have been the tools used by the authorities to increase awareness of the working environment. In future more emphasis will be put on the provision of information and counselling by the Working Environment Service system to small businesses as well as larger ones. It is intended that such businesses should have access to the expertise and environmental management systems normally only available in larger companies. The role of the Labour Inspectorate in relation to businesses will continue to be that of a supervisory authority.

Nowadays the public authorities put considerable effort into measures aimed at offering service and flexibility. This depends on a substantial development away from the traditional command structure with many hierarchical levels to a more co-operative system capable of identifying the important themes to be addressed and of taking the necessary decisions. A range of initiatives are involved, including some concerned with safety work. The feedback from this work is of a variable nature. In attempts to deal with the problems of repetitive work in industry, for example, one of the lessons learnt has been that worthwhile modifications require resources and an open approach, and that changes are all too apt to have a superficial effect. Within the public sector it is also abundantly evident that the need for greater flexibility and communication at work puts many employees in situations in which they find it difficult to adapt their ways of thinking to the demands of the new work.

A picture of Danish experience and attitudes emerged from a comparative analysis of data collected from middle managers in the 42 countries involved in a study in 1982. The distinctive features of the Danish tradition were identified as tendencies to be:

- extremely democratic
- well-developed in terms of organisational communication
- positive towards managers and employees
- independent in relation to social systems and in terms of the capability to take on assignments
- reactive to information rather than instruction
- relatively committed to health issues rather than being occupied with wealth and social esteem.

The significance of this in terms of health strategy is that the Danish tradition is well adapted to the dissemination of subject knowledge and a creative contribution by employers. In reality, however, risk has been most often defined by the supervisory activities of the authorities, and the working environment strategy of companies has followed from this rather than through participative initiatives that might have been expected from the tendencies identified above.

An unstimulating and inadequate working environment is typically the result of the scientific and bureaucratic search for efficiency, and this is entirely at odds with the fact that current values favour a high level of input and innovation in work and the working environment. Today companies and organisations are looking for something more tailored and more stimulating than the old-fashioned approach to the working environment. Some of the ideas that are current in this respect are also reminiscent of the discussions on job-satisfaction popular in previous decades. In general people would like their employment to be intrinsically motivating.

Conclusions

As a social strategy, developments in the working environment field are characterised by the fact that the Labour Inspectorate is in the forefront of all significant initiatives, and that the organisations in the labour market have supported this state of affairs. The authorities' contribution is typified by relative freedom of action but little latitude in terms of sanctions. A powerful factor has been the moral legitimacy which has boosted support for enforcement amongst the organisations in the field.

The official initiatives have also drawn strength from the cultural tradition which is influenced to a relatively high degree by considerations of health, democracy and co-operation, and which has promoted welfare at work without all-pervading supervisory official control. Against this background, the priorities in respect of advisory initiatives must be to expand the OHS system, and to control and direct the quality of consultation. At the same time, a more creative and committed approach from the companies themselves is desirable. Creative organisation of the working environment, on a basis supported by the companies' employees themselves, will be an important theme for the years to come.

An important factor in the working environment debate has been an understanding of the fact that society - that is to say taxpayers - have paid more than 95% of the costs of injuries resulting from a poor working environment. This has emphasised the moral/ethical aspect as well as a financial entitlement to participate in matters affecting work. There is no immediate prospect of this state of affairs coming to an end, and it will continue to lend the working environment debate a particular urgency in Denmark.

1 The context of national occupational health and safety policies in Finland

Economic structure

Finland has one of the largest territories in Europe, covering 338,000 square kilometres. It has a population of 5.1 millions and the average density is 16 inhabitants per square kilometre. It is the northernmost European country with 0.4% of the global GNP and 0.1% of the global workforce. Within the EU it represents 1.8% of the GNP and 1.85% of the workforce.

Within Finland, a 5% share of the GNP and 9% of the workforce comes from agriculture and forestry, 33% of the GNP and 27% of the workforce from manufacturing industry, and 62% of the GNP and 64% of the workforce from services.

Industrial production is strongly dominated by the paper-pulp, wood and metal industries; these are expected to continue to be the strongest industries in the future. The development of high-tech products in various sectors has been rapid and growth in electronics has been particularly active in recent years. The focus is on a large number of small enterprises, with almost 99% of 123,000 enterprises employing less than 100 workers and less than 250 (0.2%) of companies employing more than 500. Large numbers of small and medium-sized enterprises (SMEs) operate, particularly in manufacturing (19%), construction (11%) and trade (36%).

The service sector employs two thirds of the workforce; it grew by about 10% during the 1980s. About 50% of service employees work in the public sector and the rest in private services. Employment is highest in public administration and defence, education and research, health services and social security in the public sector, and trade in the private sector.

Agriculture produces 2.6% of the GNP and employs 7.2% of the workforce. About 50%

of farms have livestock: dairy production, pigs and poultry. Forestry constitutes an important source of income for Finnish farmers. The average size of farms has grown rapidly over the last 20 years, although the number of very large farms (with over 100ha of land) is low (0.3%) Most farms are family enterprises employing only one to two family members. Finland is self-sufficient in almost all agricultural products and over-production causes problems in both milk and crops.

Traditionally the public sector has been important in a small country where private capital has been insufficient to establish the necessary infrastructures. About one quarter of the workforce and 42–43% of the GNP comes from the public sector. As the economic recession of the early 1990s particularly affected the private sector, this had the effect of increasing the share in the economy of the public sector, although its overall size decreased. About 530,000 people (24% of the workforce) are now employed in the public sector, while one million people (40% of the workforce) work in private enterprises. The rest are employed in state-owned or municipal enterprises, or in certain non-governmental, non-profit organisations. The private sector is made up of 50% of the workforce in services, the whole of the workforce in agriculture and the majority of industrial workers.

The labour market

The workforce comprises 2.5 million people of whom about 2.0 million are at present employed and 0.5 million (20%) are unemployed. Labour participation rates are on average high (77.2%), particularly the participation rate of female workers (73.1%). The participation of men over 55 years of age, on the other hand, is low, reflecting the high incidence of work disability.

The unemployment rates were low (3.5–5.5%) for the whole of the 1980s, but structural changes in the economy, technological development in production, national policy, economic recession and the collapse of markets in Eastern Europe caused a rapid increase of unemployment so that it has reached 20% in three years. Unemployment among young workers under 24 is as high as 30%. Parallel with the growth of unemployment, short-term work contracts have increased; in 1991, for example, 55% of people recruited entered short-term employment. Part-time employment has also grown, from 2% to about 8% currently. Paradoxically, at the same time there has also been a slight growth in overtime work.

The demographic trends of the Finnish workforce are dominated by a rapid increase in the average age of working people. This is exacerbated by the shortage of young labour resulting from the low birth rates that have prevailed for 20 to 30 years. Longer training periods and high unemployment rates among young people have lowered participation rates to 27.5% amongst the 15–19 age-group. This will further aggravate the ageing of the workforce.

High participation by women is possible because of high educational status, equality regulations, well-organised day-care for children and 263 days' well-compensated maternity leave. A number of special legal provisions protect female workers, such as limitations regarding night work and special provisions for the protection of reproductive health at work. Despite the strong emphasis given to equality issues, the salary levels of women are still lower than those of men, and job segregation between the sexes still exists for 46% of employees.

The level of education among older age-groups corresponds to the European average, but that of young people is amongst the highest in Europe. As many as 83% of the 25–34 age-group have either university or other post-school qualifications.

Occupational structures are changing rapidly along with changes in economic structures and with rising educational levels. Employees in the industrial and primary production sectors are decreasing at the rate of 5–10% per ten years, while experts and those in white-collar and service occupations are increasing in number. The general trend is that the musculo-manual occupations are gradually disappearing, while computerised expert occupations increase in numbers.

The number of migrant workers is extremely low in Finland, less than 1% of the total workforce, but it is expected to increase in the future.

Industrial relations

As with the other Nordic countries, Finland is a consensus-society where tripartite collaboration between government, employers and employees has developed well and expanded far beyond the limitations of traditional salary bargaining. This has made the development of smooth and effective industrial relations an issue of utmost importance affecting not only labour issues, but also other aspects of society. Collective agreements with extensive economic, social policy, salary and occupational safety and health aspects have been significant instruments of societal development. Such agreements cover almost 100% of Finnish employees.

Related to the sense of the importance of industrial relations and social dialogue are the rates of unionisation of workers, employers and the self-employed. These are exceptionally high, at about 90% of employees and nearly 100% of active farmers, as well as the vast majority of the employers including private enterprises, the public sector and SMEs.

There are well-established structures for industrial relations and collaboration, such as Works Councils, tripartite committees and working groups. The official organs for collaboration in occupational safety and health at the national level are bodies established by legislation such as the advisory committees for occupational safety and health and for the occupational health services. Collaborative committees and occupational safety committees, which also have their status, composition and functions stipulated in the legislation, are the respective plant- and company-level organisations.

Values and philosophy in relation to health and safety

The value base of occupational safety and health in Finland reflects the wide consensus on the central position of work in the citizen's life, collaborative principles in the policies of various partners and the appreciation of occupational safety and health as an asset and instrument for better product quality and better productivity, as well as better quality of life. The Government's values are spelt out in the constitution of the Republic that emphasises that labour enjoys the special protection of the State. That principle is detailed in the occupational safety and health legislation which emphasises the priority of health and safety at work over material values. The employers emphasise both the humanitarian value of occupational safety and health, and the economic and operational benefits of a safe and healthy work environment. The trade unions and Farmers' Union stress safety and health at work as one of the basic rights of citizens as laid down by the constitution.

The European dimension

As a member of the EEA, from 1st January 1994 and the European Union from 1st July 1995, Finland has implemented the European Union Directives including the Framework Directive (391/89/EEC) and its daughter directives. The national legislation and regulations have been adjusted to the EU regulations. There were no major difficulties and the practical impact remains relatively low. The European dimension has been understood to be mainly an economic one and care has been taken to preserve the standards of occupational safety and health that Finland has achieved regardless of the pressures of international markets.

2 Occupational health and safety policies and structures

Legislation

Finland's tradition of over 100 years of occupational safety and health legislation reflects the principles of prevention, protection, participation and the primary responsibility of the employer in occupational safety and health. The major legal instrument that regulates the substantive content of occupational safety and health is the Labour Protection Act, which was first passed in 1958 and has been amended several times, most recently in 1993. The Act stipulates the basic responsibilities of the employer and the rights and responsibilities of the worker and the authorities, such as the Government's occupational safety and health authority, for the inspection of compliance. The Act is supplemented with several specific pieces of legislation regulating, for example, work by women and young people, as well as numerous other regulations.

In addition, the Occupational Health Services Act 1978 obliges the employer to organise and finance such services for all workers using the expertise of competent health personnel.

This substantive legislation is implemented with the help of two instrumental legislative provisions:

- the Labour Protection Administration Act 1993 stipulating the organisation, tasks, functions and authority of occupational safety and health administration
- the Labour Protection Supervision Act 1973 and subsequent amendments that stipulates the organisation, functions, authority and tasks of occupational safety and health inspection; the rights, duties and tasks of occupational safety committees at the company level; and the rights and obligations of safety representatives.

Typical of these legal instruments is that practical decision-making on occupational safety and health, and the implementation as well as follow-up and inspection of occupational safety and health activities are all carried out, where possible, by tripartite collaboration.

Occupational health and safety structures

The organisation of occupational safety and health is based on the principle that the workplace is the key focus of occupational safety and health activity and it is advised, supported, controlled and monitored by the authorities, expert institutions and social partners. The principal policy-making body is the Ministry of Labour with its two divisions responsible for various aspects of occupational safety and health: the Division of Work Environment and the Division of Occupational Safety and Health. The Ministry of Social Affairs and Health controls the professional activities of the occupational health services and the legislation on workmen's compensation. Numerous other bodies, particularly the Institute of Occupational Health with its six regional institutes and state Technical Research Centre, provide expert advisory services, research, training and information support, while the Centre for Occupational Safety provides training and information for plant-level organisation, workers' representatives and members of the safety committees.

Finland has a special Work Environment Fund to provide financial support for practice-oriented applied research in occupational safety and health, training, information and for the application of research results in practice. The social partners have their own expert resources and special officers or units for occupational safety and health issues.

Control and inspection

There are 11 labour protection district offices covering the country working in occupational safety and health inspection with a staff of 423. Centralised machinery and product control inspections, market surveillance inspections, workplace inspections, special inspections and advance inspection of the plans for industrial facilities are the main activities. About 30% of workplaces with recognised occupational health and safety risks are inspected annually, while only about 7–15% of less risky workplaces are covered. The average density of inspectors is about 2 per 10,000 workers, and about 70% of their capacity is used in high-risk sectors employing about on average 580,000 workers. About 70% of inspectors have at least middle-level (technical) education and they have been trained by the Ministry of Labour in occupational safety and health issues. The inspectors have the right under the Supervision of Labour Protection Act to enter any workplace without advance notification.

Inspectors are authorised to order changes relating to occupational safety and health, issue penalty fines with deadlines, stop the work or close the whole plant, as well as to give advice for the improvement of occupational safety and health in the company. About 60,000 instructions for improvements were given annually during the 1980s, but only 85 of these (0.14%) were binding requirements for the employer to make corrections. Since 1973 less than 1% of all inspections has led to prosecution.

Occupational health services

The occupational health services (OHS) were created by a 1978 Act which obliges the employer to organise services for all employees regardless of the size of the company, sector of economy, employer-employee relationship or occupations. The compulsory aspect of the services are purely preventive, but treatment services are also available. The approach of the OHS is expected to be comprehensive, multidisciplinary, to use up-to-date validated methods, competent staff and focus activities on workers, the work environment and work organisation. To meet these objectives, principles of participation and collaboration need to be developed.

The employer is compensated for up to 50% of the cost of OHS from the sickness insurance funds. The organisational model for service delivery can be freely chosen by the employer after consultation with the safety committee or the workers' representative. The main service provision models are in-plant service, municipal health centres, private health centres and group services. The coverage of the Finnish OHS is high, at present 90% of employees and 85% of the total workforce. The self-employed are entitled to participate through municipal health centres which have a legal obligation to provide services if requested.

There are about 4,850 people employed on a full-time or part-time basis in the OHS, such as OH doctors and nurses, physiotherapists, psychologists and auxiliary personnel. They are able to collaborate with about 12,000 safety officers at the plant level and with about 60,000 members of the safety committees.

The total budget of the OHS is 180 million ECU, i.e. 109 ECU per worker served by OHS (0.2% of the GDP). The employer's compensation from the Sickness Insurance Fund is 99 million ECU. The money for the compensation is collected from employers and workers in the form of a social security charge. Theoretically the OHSs are well integrated with the safety activities in the workplace and with primary health care in the community, but in practice such links could be better and the exchange of information could be more active.

Information and assistance

There are numerous legal provisions obliging the employer to provide information to the authorities and to workers, including the duty to inform the workers on hazards at work and their avoidance. There is also the obligation on the part of the occupational safety and health authorities to provide information on regulations to employers and workers, and the obligation for the employer to provide inspectors with all the information they need for inspection. The workers and their representatives must inform the employer of any hazards observed at work, and immediately notify the employer if they interrupt hazardous work or refuse to carry out a dangerous job.

Several types of advisory information guidelines, booklets, journals, publications, instructions, training and educational material are provided and distributed by the occupational safety and health authorities, the Institute of Occupational Health and the Centre for Occupational Safety, as well as the employers' confederations and the trade unions. This material amounts to about 30–40 pages of written information for every worker per year.

Training

The Ministry of Labour, Ministry of Social Affairs and Health, Institute of Occupational Health, Centre for Occupational Safety, trade unions, employers' institutions and several professional associations organise complementary training in special courses or seminars for workers, their representatives, safety officers, occupational health experts, managers and employers. A rough estimate of the total volume of such training per year is about 20,000–30,000 person-days, involving about 15,000–20,000 people, i.e. on average each person active in occupational safety and health at the workplace level will participate in training once in three years. The turnover of such personnel, however, reduces the rate of training.

The basic curricula for doctors, nurses, engineers and some other expert categories includes a short course which provides basic information on the objectives and activities in occupational safety and health. Specialist training is provided for occupational health doctors. There are several courses for occupational health nurses, physiotherapists, psychologists and a three-year specialist training for occupational hygienists with an examination and a certificate given by the Institute of Occupational Health.

The primary school and vocational school curricula have so far contained only a minimal amount of occupational safety and health training. A new strategy for vocational training, however, contains a substantial element of occupational safety and health as an integrated part of the curricula for all occupations and professions.

Economic incentives

The economic aspect of occupational safety and health has been the subject of discussion in Finland, particularly in the light of economic recession. One notable issue is the economic appraisal of the direct and indirect economic loss caused by occupational safety and health hazards. Also, the consequences of occupational safety and health regulations and standards are being analysed in the process of preparing new regulations. This appraisal has revealed that the total loss through injury and disease (including all calculable and estimated indirect losses) may be as high as 5–15% of the GNP. The occupational safety and health authorities, together with the research institutions, have started a programme for the development of methodologies for economic analysis. The focus has been on three different contexts:

- when occupational safety and health regulations and measures have a clear positive economic impact when measured with any parameter
- when the impact is negative at the individual enterprise level but positive at the level of the national economy
- when the impact is negative at the national level but the nature of the problem still calls for action.

For the purpose of the calculation of the economic impact at enterprise level, two tools have been developed:

- the software for calculating the costs of occupational accidents and the benefits of safety investment
- a model for personnel economy accounting systems.

Financial incentives for the promotion of occupational safety and health include:

- risk-related insurance premiums
- grants for research, training, information and the application of research results from the Work Environment Fund
- financial support for trade unions for training occupational safety and health personnel
- compensation for the costs of the occupational health services
- the advisory, research and expert support and consultations provided for workplaces by the Institute of Occupational Health and its regional institutes.

Research

Research activities in occupational safety and health are comparatively well established. The key institutions are:

- the Institute of Occupational Health with a broad spectrum of research activities covering several disciplines relevant to occupational safety and health
- the State Technical Research Centre with technically-oriented research
- universities, each with a relatively limited scope of research, but as a whole constituting a multidisciplinary research resource covering all relevant fields of occupational safety and health research.

About 10–15 institutions are active in the field of occupational safety and health research and the total research personnel consists of 520 people, about half of them working in the Institute of Occupational Health.

The vast majority of research is practice-oriented applied research for which an active research training is required; the seniority level of the research personnel is high. The professional scope of the expertise covers most of the areas relevant for occupational safety and health. Some shortages prevail, however, in the fields of social sciences and the economic aspects of occupational safety and health. The total budget is about 25 million ECU, constituting about 0.3% of the GNP, 1.5% of the total national research and development budget, and 3.9% of the Government budget for research and development.

A recent Ministry of Labour working group listed eight priority areas for research which should be promoted in the next few years. They were:

- the planning of safe technologies and working methods
- psycho-social aspects at work and well-functioning work organisations
- the causes and mechanisms of work-related diseases
- health hazards and their prevention
- the causes and prevention of occupational accidents and major hazards
- the development of occupational safety organisations
- the development of occupational health services
- the effects of occupational safety and health on national and corporate economies.

There are several mechanisms for distributing the research results to the users, but certain difficulties have been identified in the application of the specific research results in practice. The process tends to be dependent on the training of experts and the development of validated service methods rather than the direct application of individual research results. This makes the role of expert practitioners, such as occupational health service staff, safety officers and safety representatives, crucial.

Occupational health insurance

Insurance for occupational accidents and occupational diseases must be arranged by the employer for all employees. The self-employed have the same obligation and accident insurance thus covers a total of 2.4 million people. The Labour Accident Insurance Act has priority over other social security provisions and the level of compensation is about 100% for short-term disability and about 80% for long-term disability. Money is provided for care and treatment, rehabilitation, work disability, handicap and inconvenience, pensions for survivors, special medical examinations, prosthesis and aids. Vocational training and replacement costs are also covered. The insurance premiums partly contain prevention-oriented incentives by rating the premiums for larger companies according to the accidents experienced in the company. A similar procedure operates for small enterprises where the level of premium is calculated according to the average risk of the sector.

Monitoring

There are four principal kinds of monitoring of working conditions, occupational health and safety, and the health of workers:

1 The monitoring of the work environment and exposures at the workplace is carried out by the OHS teams. The enterprise must keep the monitoring data and make it available on request for inspection by the occupational safety and health inspector, the safety committee and the safety representative.

2 The biological and medical monitoring of workers' health is compulsory for all workers in especially demanding or hazardous jobs, with the frequency of monitoring varying according to the risks.

3 The specific monitoring of branches or sectors of industries and other economic activities that is carried out as special projects. This monitoring may also be problem-oriented, such as the monitoring of chemical exposures in the work environment. It is based on special surveys that may be repeated

periodically.

4 Monitoring at the national level by using national registers and data systems. The most important data systems available are :

- the national statistics providing occupational data
- several morbidity registers including hospital discharge data, cancer registry, registry of congenital anomalies, etc.
- the national registry of occupational and work-related diseases
- the national registry of occupational accidents
- the registry of occupational hygiene measurements
- the registry of exposure of workers to carcinogenic substances
- the registry of asbestos-exposed workers
- the registry of chemical safety data sheets for workplace and consumer products.

The linkage of such data systems has been possible, but difficulties are encountered because of stringent regulations on data protection. Such national data systems have been most instrumental not only for monitoring the national occupational safety and health situation, but also for focusing control and regulatory actions, and for epidemiological research.

The bodies carrying out this monitoring are the Ministry of Labour, Ministry of Social Affairs and Health, Statistics Finland, Institute of Occupational Health, Association of Accident Insurance Institutions, Social Security Institution, National Cancer Registry and certain research bodies.

Emerging issues

Due to rapid changes in the economic structure, growing internationalisation and European integration, dynamic developments in working life are predicted. There are a number of major issues which it has been recognised need a nationwide response. These include:

1 The adaptation of the occupational safety and health regulations and inspection practices to EU Directives and policies.

2 Structural changes in the economy, division of work, occupational structures and exposure as a result of the integration of new technologies and the introduction of new production methods. As a consequence of these changes, it will be necessary to introduce new training methods and working practices including those for occupational safety and health.

3 Rapid demographic changes in the working population due to the ageing of the post-war baby boom generation.

4 The introduction and re-introduction of vocational skills and occupational safety and health knowledge and practices for presently unemployed people who may get employment later.

5 The introduction of quality management systems into occupational safety and health activities.

6 The special occupational health problems related to chronic non-communicable diseases affecting the working capacity. The maintenance of working capacity and prevention of work disability, development of psycho-social working conditions and well-functioning work organisations, prevention of hazards of carcinogenic exposures, reproductive health hazards, allergens and musculoskeletal disorders.

7 The stronger integration of enterprise-level occupational safety and health practices with production lines, with other services and with community services for health care and environmental protection This requires further training of occupational safety and health personnel and the establishment of collaborative links both within enterprises and with the community.

8 In research, more predictive risk assessment methods will be developed in order to give support for primary preventive measures in various problem areas.

9 Growing international collaboration is envisaged at the global and European level. Collaboration with the developing countries and with the countries of Central and Eastern Europe is needed, especially those neighbouring Finland. The Nordic collaboration will be continued and further strengthened on the basis of the Nordic Convention on Occupational Health and Safety.

3 Health and safety outputs

Occupational accidents

The total number of accidents and fatal accidents have been declining since 1985, particularly for occupational accidents. In 1991 the average rate of accidents leading to three days or longer absenteeism was 52 per 1,000 workers or 26 cases per million working hours. This was 27% lower for all types of accidents and about 50% lower for fatalities in 1991 compared with 1977. The rates of all accidents which were registered by the insurance system were 38 per million hours, indicating that a substantial part (about 40%) of all accidents result in less than three days absenteeism.

The average risk of fatal accidents in 1991 was about 0.036 per million hours and about one in three of the fatalities were commuting accidents. The highest fatality risk was registered in mining and quarrying, and the lowest in the food industry. About 0.07% of all accidents were fatal and 0.48% caused permanent disability.

Occupational diseases

The total number of registered occupational diseases in 1992 was 8,034 cases, giving an average rate of 37 per 10,000 employed with a 42-fold variation between the highest and lowest-risk occupations. In 1994 the number of cases was 6,672. In 1992 one third of occupational diseases were caused by repetitive tasks or monotonous work (strain injury), about 15% by noise, 15% were dermatoses and about 30% were caused by numerous individual factors, amongst which the occupational respiratory allergens were the most prevalent and steadily growing factor. While the number of cases of skin diseases and noise-induced hearing loss have remained relatively constant since 1976, those caused by repetitive or monotonous work have increased between 1986 and 1991.

Absenteeism and work incapacity

Finnish absenteeism rates are comparatively low in general and particularly the sickness absenteeism rates, causing 4.7% work-time loss in services and a 4.6% loss in industry. On the other hand, both the longer-term disability and disability pension rates are exceptionally high. The prevalence of work disability amongst workers over 55 years makes participation rates amongst this group comparatively low (26%). The rates of sickness absenteeism have been slightly declining in the 1990s, while the number of disability pensions still grows.

Work environment surveys

Several surveys have been carried out and will be repeated periodically for the assessment of the present status of work and the work environment. A survey of the working conditions of employees is made periodically in connection with the workforce survey by Statistics Finland. Interviews, questionnaires and analysis of available statistical data provide the source of information for such surveys. The overall conclusion of such surveys, based on both objective and subjective assessment, indicates the general improvement in the physio-chemical occupational safety and health situation compared with 10 years ago. Increase in the psychological work-load, time pressure and requirements for increased productivity and quality of products are causing more prominent occupational safety and health problems, particularly when taking account of the rapid ageing of the workforce.

4 Assessment of occupational health and safety policies

The policy assessments of the 14 bodies of decision-makers, social partners and various participants in occupational health give relatively positive results and indicate an interest in the further development of occupational health and safety. Policy priorities cover the effective response to the rapid change in the economic structure and new technologies, response to problems of the ageing workforce, work disability and the effective adjustment of the occupational health and safety system to EU occupational safety and health policies and regulations. The Finnish occupational safety and health policy will continue its development on the basis of national traditions and experience - carrying out active international collaboration, particularly with the Nordic countries and with the EU - by making effective use of cumulated experience and utilising opportunities to facilitate general socio-economic development with the help of an occupational health approach.

The strengths mentioned by most of the experts interviewed were the legislation-based infrastructures run by competent staff and supported by research institutions. The weaknesses, on the other hand, were the holes in the coverage of services in certain sectors and the inequality associated with this; also the division of leadership and responsibility between two Ministries. The social partners recognised the problems of the ageing workforce and high unemployment rates. Priorities were similar among different participants. They emphasised the need to develop the service infrastructures, quality of services and collaboration, and to consider the ageing trends of the workforce. It was felt on the one hand that the impact of the economy might cause a lowering of the priority of occupational safety and health programmes, but, on the other, that it might focus the activities on the most important targets.

The main problems for the next ten years were seen to be multiple, including organisational objectives, the further development of occupational health services, training systems and the quality management of services. Substantial objectives are the maintenance of the working capacity, prevention of traditional occupational safety and health hazards, and increasing possibilities for participation.

The impact of the EU was expected to be minor and mainly positive, while some questions were raised about sustaining the achieved level of occupational safety and health within the EU system. On the other hand, the stabilising effect and the promotion of occupational safety and health throughout the region were seen as positive impacts of the EU.

1 The context of national occupational health and safety policies in France

Economic structure

Although France has kept a significant share of its activities to the primary sector for longer than other mainland northern European countries and the United Kingdom, today France has a manufacturing sector whose structure is very close to theirs and which is evolving in the same way.

In France, more than half of those employed work in small or very small businesses (under 50 employees), representing 97% of all businesses; 24% of employed people work in firms of 200 or more, which represents only 0.6% of the total. This is significant because committees for health and safety and for working conditions only exist in companies that have more than 50 employees, and there are no health and safety representatives in companies of under 10 employees.

The labour market

The crisis affecting all industrialised countries has meant a very high level of unemployment in France, characterised, amongst other things, by long-term unemployment, and unemployment among young people and women in particular. The active foreign population has stabilised, there are less people working in both the oldest and youngest age-groups, and part-time work, which was traditionally underdeveloped in France, has seen a remarkable growth, in parallel with the growth in atypical forms of employment.

Industrial relations

The industrial relations system is characterised by the extreme weakness of the unions, the level of union membership being probably under 10% of workers, concentrated mainly in large companies in the public sector. Negotiations, which for a long time centred at industrial sector level, are now increasingly taking place at company level, and are only rarely concerned with questions of health or safety.

This weakness on the part of the unions has its impact on the proper functioning of the representative bodies. Even though the legislation provides palliatives for this deficiency, it seems that only 65% of companies which ought to have a health, safety and working conditions committee actually have one. Small companies, and particularly the smallest of these who together employ over 3.6 million workers, have no representative structure whatever and unionisation is entirely absent.

The European dimension

In spite of this general context, the public authorities and the social partners have made a great effort to implant the new European standards. In 1992 there was significant movement on the problems of health and safety. The Caisse Nationale d'Assurance Maladie (CNAM) (National Health Insurance Fund) and the INRS (National Institute for Research and Safety for the Prevention of Industrial Accidents and Occupational Diseases) created a public interest group, EUROGIP, in 1991 whose aim was to co-ordinate and develop at European level the activities of the social security bodies in the area of hygiene, safety and health.

2 Occupational health and safety policies and structures

Legislation

The development of health and safety legislation is based first and foremost on the legislature's will to protect particularly vulnerable groups (children, women) which are often poorly represented by the trade unions. As a general rule, every time it can the legislature intervenes subsequent to an agreement between the social partners so that rulings put together in the form of a labour agreement can be extended to the body of the employed population which is likely to be concerned. Quite often, in the area of health and safety, the legislature has been the one to take the inititative in the face of failings or insufficiency by the social partners. The legislature tends to replace corrective policies with those fostering prevention and the development of employee and employee representative participation, doing this also by involving the latter in the social security agencies and public bodies entrusted with the implementation of such preventive policies.

Health and safety provisions are found in the Labour Code. They date from the late nineteenth century when the Industrial Establishments Act 1893 gave employers a duty to provide clean and safe working conditions. The main instrument establishing preventive services dates from an Act of 1946 which introduced occupational health services. Its coverage has gradually been extended in various new laws since that time.

Many public employees do not fall under the scope of the Labour Code and separate provisions have had to be introduced to extend similar protection to different groups within the public sector. Legislation on worker representation in health and safety was originally introduced after the Second World War by a Decree in 1947. Several reforms followed in subsequent decades, the most significant in 1982 which gave workers' representatives greater protection and paid release for training as well as lowering the threshold requiring firms to establish joint health and safety committees (CHSCTs) to those with more than 50 employees.

The Act of 6 December 1976 played an important part in defining a general prevention policy by setting up a tripartite Industrial Injury Prevention Board that must be consulted on all new draft legislation and regulations regarding prevention.

The Prevention of Occupational Risks Act 31 December 1991 incorporated the EU Framework Directive 89/391/EEC into French law as well as transposing the provisions of other EU Directives. It amends the provisions of the Labour Code but therefore excludes from its coverage central government civil servants and local and regional authorities.

Control and inspection

Great efforts have been made by the authorities and the social partners to develop prevention policy. Examples of this include the creation of an occupational hazards monitor (Observatoire des Risques Professionels), the introduction of targeted campaigns for the Labour Inspectorate and the increase in the number of company medical officers.

The Labour Inspectorate in France has extensive powers since its authority encompasses the control and application of the labour regulations, conciliation, arbitration and mediation procedures, as well as participation in the implementation of employment policy. Since January 1994, the organisation of labour inspectors of transport and agriculture have amalgamated with the organisation of industrial labour inspectors. The excellent training received by labour inspectors and controllers is, however, not enough to make up for their lack of numbers. The extent of the tasks entrusted to them and the often contradictory roles they have to play have caused a certain amount of tension within the profession. If the numbers of inspectors are compared with the numbers of establishments to be inspected, it is clear that there is only one inspector per 1,050 establishments and per 10,700 employees. Furthermore, the economic context makes it more difficult to apply stringently the legislation governing health and safety.

Offences and penalties

Breaches of the health and safety regulations represented over 40% of offences acted upon by the Labour Inspectorate in 1990. In 8% of the cases there were breaches of regulations by employers in relation to company medical services. In one-third of the cases (7,500 sentences) the courts applied penalties in excess of the minimum.

Occupational health services

Between 1982 and 1989, a steady decline could be observed in the numbers of company medical officers. Since that time, however, their numbers have risen significantly, which means that the number of employees under their care has also increased (85% of employees). The numbers of medical officers are proportionate to workplace size (i.e.

number of employees) and the great majority of companies make use of inter-company medical services.

Company medical officers are responsible, apart from an annual medical visit to check on the individual health of employees, for reserving one-third of their time ('the third') for analysing jobs as well as health and safety conditions in the company, and for advising management and the health and safety committee. They are also expected to be consulted on the introduction of new production techniques. Since 1988, the company medical officer has been required to present an annual plan of his/her activities, indicating studies to be undertaken and visits to workplaces to be made. Medical examinations can take place every two years rather than annually, allowing the doctor to devote more time to analysing work situations and joining with other consultants for the development of a multidisciplinary approach to occupational health (involving 100,000 paid employees). The specific areas of supervision that apply to temporary workers have also been stipulated by decree since 1991.

Company doctors are now better trained, but less than 50% of them are involved full time and only doctors who are fully integrated into a company can really perform effectively their designated role to improve working conditions. The cost of the medical service is met by employers as a whole.

Information and assistance

The rules about informing and educating paid employees in relation to health and safety are important, but their non-application is regularly sanctioned. Some of these rules provide for information for individual workers concerning the risks they run and the protective measures they should take; these rules apply equally to temporary workers and subcontractors. To this information should be added, where necessary, specific training measures. In addition, a certain amount of information on health and safety should be given to the health and safety committee and to the Labour Inspectorate. The employer is expected to present an annual action plan to the health and safety committee accompanied by a budget to cover the plan. The members of the committee are entitled (under 1991 legislation) to a five-day training course for companies employing over 300 people, or a three-day course for companies employing fewer than this number. The cost of such training courses is borne by the employer. It should be added that the National Institute for Research and Health sets aside over 35% of its budget for the information and education of paid employees and students of technical schools.

In the building and public works sector a special organisation has been in place since 1947: the Organisme Professionnel de Prévention du Bâtiment et des Travaux Publics (OPPBTP). This is a professional representative body for the prevention of accidents in the construction industry whose activities are organised to provide advice and training to companies in this sector. These companies were regulated in 1982 under legislation for companies with over 300 employees; it was then that the health and safety committees took over some of the functions of the OPPBTP. An Act passed in 1991 extended the legislation to companies of all sizes in this sector; it introduced special regulations governing matters of health and safety on long-term building sites and the appointment of a site co-ordinator responsible for health and safety where a number of companies are operating on the same site. The OPPBTP continues to provide information, training and advice.

Occupational health insurance

CNAM has the task of administering the insurance contributions in respect of industrial accidents and occupational diseases. The Regional Technical Centres (CTR) which are equally representative of employers and employees, are consultative bodies associated with the regional health insurance funds; they classify companies according to category of occupational risks and determine the level of contributions they should make. The system they operate is very complex and varies with the size of the company. Large companies are subject to a fixed rate; medium-sized companies pay a fixed rate plus a sectorial rate; small companies are subject to a rate calculated for their sector. Since 1987 the CNAM has pursued a more active policy of prevention by signing 'objective contracts'; these reward improvements and good practice by offering bonuses and penalties in much the same way as motor insurance operates. The system for the agriculture sector is managed by the Mutualité Sociale Agricole (MSA) on a similar basis.

Research

Research on the subject of health and safety is officially the responsibility of the National Institute for Research and Safety for the Prevention of Industrial Accidents (Institut National de Recherche et de Sécurité - INRS). This Institute, which is dependent on the CNAM, and whose budget is derived from company contributions, carries out basic and applied research, the results of which are widely disseminated to employers and government authorities.

The INRS does not have a monopoly of the research in this field, however. Research into health, safety and occupational hazards is also carried out in research laboratories associated with universities and at the Centre National de la Recherche Scientifique (CNRS) as well as by INSERM. In addition to the work of these public research teams, there is the research carried out within large companies and that conducted by specialist institutes for the main trade union organisations. In the absence of a special survey it is impossible at the moment to evaluate the research budget or the number of researchers who devote all or part of their time to research in health and safety.

3 Health and safety outputs

Occupational accidents

Having declined steadily until 1988, industrial accidents, including fatal accidents, have seen a significant re-emergence. This increase was judged sufficiently alarming for the public authorities to decide on the creation of an occupational hazard monitor (1990) in addition to inspection campaigns and special building measures.

A detailed analysis of accidents at work shows a concentration of accidents in certain sectors (building and public works, forestry, mechanical engineering, auto repairs). It highlights the risks of accidents for workers in small companies, the least-skilled and the youngest workers and those whose status is insecure. Foreign workers in sectors at risk also have more accidents than nationals. To the accidents at work must be added accidents in transit, the number of which is also growing rapidly. For 1991, the cost of accidents at work rose to 17 billion francs, or 2,607,361,963 ECU.

Occupational diseases

Occupational diseases have remained stable over the medium and long term. 4,100 cases are identified annually and three diseases represent 50% of those diagnosed: periarticular disorders (of the joints) and disorders due to noise and asbestos. Certain sectors have a high rate of both accidents and occupational disease.

4 Assessment of occupational health and safety policies

For all the respondents consulted, the economic context was seen to be particularly significant in the assessment of the policies put forward. They emphasised, for instance, the intensification of work associated with the reorganisation of production methods, the growth of both shift work and casual jobs. Whilst working conditions in general are deteriorating - as confirmed by surveys conducted by the Ministry of Labour and by the National Statistics Institute - the extent of unemployment makes it more difficult to apply sanctions that are likely to compromise jobs, and for workers who are subject to poor working conditions to be more demanding.

The measures for prevention and control, officially remarkable in many respects, are severely handicapped by the extreme weakness of the unions and by the insufficient numbers of Labour Inspectorate personnel. Some union representatives have reservations regarding the company medical services measure which makes the doctor a company employee and they deplore the small amount of time devoted by company doctors to analysing work situations. All the respondents hope to see more efforts made in the area of research, but they generally appreciate the quality of the research done by the public bodies. However, the absence of any research conducted in certain sectors, particularly the service sector, must be deplored.

The problem of insurance contributions against industrial accidents or occupational diseases recurs regularly in discussions on prevention policies. The grouping of contributions for small companies has been denounced for its adverse effects in the area of prevention. All the proposed measures intended to change the system meet with opposition from the representatives of small companies.

Small companies constitute a major difficulty in implementing effective prevention policies. Inspections at such companies are less frequent, worker representation is non-existent and intervention by the company doctor very sporadic. No practical solution has so far been found for the problems of this group.

Finally, the State as employer is revealed as being particularly in default in occupational safety and health.

In general, for the organisations interviewed, the most interesting innovations in recent years have been the 1982 legislation known as the 'Auroux Acts' covering recourse to expert advice, the role assigned to the health and safety committees and worker participation. The social partners have approved the recent reforms in the building sector and the positive outcomes from the European Year of Health, Hygiene and Safety. With one exception the social partners favour the development of European regulations on safety and health.

Conclusions

The major preoccupations for the future are linked to the restrictions imposed by the new forms of work organisation, the problems associated with the increase in casual and precarious work and the development of prevention policies to be implemented by small companies.

Particular areas for research to be focused are:

- jobs in the tertiary sector
- the development of instruments for the evaluation of policies
- government support.

A special effort should be made in the matter of health and safety education in secondary schools and universities as well as in the workplace.

1 The context of national occupational health and safety policies in Germany

Economic structure

In April 1991, the total number of people gainfully employed in Germany was 37,445,000, over 7 million of whom were resident in eastern Germany. By sector, there were:

- 1,575,000 people in agriculture, forestry and fisheries (the primary sector)
- 15,350,000 in manufacturing (the secondary sector)
- 6,666,000 people in wholesale and retailing, transport and communications
- 13,854,000 in other fields of economic activity.

In May 1987, there were 2,579,297 workplaces employing a total of 26,867,165 people, of which 21,915,838 worked in the private sector. By workplace size these were:

- 5,444,927 people in 1,829,890 firms with 1–9 employees
- 2,056,748 people in 156,253 firms with 10–19 employees
- 6,890,289 people in 108,353 firms with 20–49 employees
- 7,523,874 people in 3,357 businesses with workforces of 500 and over.

The labour market

Unemployment

In the first quarter of 1993, the unemployment rate averaged 8.0% of the workforce in former West Germany and 16% in former East Germany and East Berlin. The number of people involved was 2,223,132 in the western and 1,165,233 in the eastern parts of the country.

Part-time work and moonlighting

The proportion of the workforce in part-time employment increased from 11.7% in 1972 to 16.2% in 1989. Estimates of the number of people involved in the informal economy in the Federal Republic of Germany in 1984 were between 2.22 million and 4.35 million.

Age and sex of the workforce

In April 1991, a total of 1,643,000 people under 20 years of age and 3,936,000 over 55 were in gainful employment. Of the 37,445,000 people gainfully employed in April 1991, 15.57 million were women. The workforce participation rate for women is 57.1%, which is below the international average.

Employment of non-nationals

In June 1992, the number of non-nationals employed in western Germany was 2,036,154, of whom 493,874 were citizens of EC countries. Migration statistics show a net inflow of 544,979 people in 1991, 356,824 of whom were non-nationals.

Industrial relations

Freedom of association and collective bargaining are both guaranteed by Article 9, para.3 of the Basic Law. The system of collective bargaining agreements is governed by the Collective Bargaining Contracts Act (Tarifvertragsgesetz - TVG). The TVG stipulates that only recognised trade unions are entitled to negotiate collective bargaining contracts. These unions are the federation Deutscher Gewerkschaftsbund (DGB) and its member unions, plus the separate white-collar union Deutsche Angestelltengewerkschaft (DAG). On the employers' side, such agreements may be concluded by employers' federations (umbrella organisations), by single employers' associations or by individual employers.

The legal framework for the establishment of works councils is provided by the Works Constitution Act (Betriebsverfassungsgesetz - BetrVG) as amended in 1972. This applies to all workplaces with more than five full-time employees. The exemption for small firms has the effect of excluding roughly 1-2 million employees from any legally guaranteed representation of their interests. Taking all exemptions together, approximately 5-6 million employees are excluded.

All employers who appoint a company doctor and safety officer are also obliged by the 1973 Safety at Work Act (Arbeitssicherheitsgesetz - ASiG) to establish a health and safety committee.

Membership of trade unions and employers' federations

In December 1992, the participating unions in the DGB federation had a combined membership of 11,015,612. It is estimated that about 40% of employees in former West Germany are members of these or other unions, while 80% of the employers in industry and in banking and insurance are members of associations belonging to the employers' confederation, and they have approximately 90% of the region's employees working for them.

2 Occupational health and safety policies and structures

Legislation

In the Federal Republic of Germany, each individual's right to life and freedom from physical injury is guaranteed by Article 2, para. 2 of the Basic Law. The prevention of accidents and occupational health hazards must take priority over compensation. The occupational health and safety legislation which serves to perform that task, also known in German as 'work protection' (Arbeitsschutz) legislation, can be divided into three areas to reflect the instititutions that are responsible for it:

- Government 'work protection'
- autonomous legislation consisting of the accident prevention laws issued by the mutual indemnity associations (MIAs) as provided for by the pre-war National Insurance Code (Reichsversicherungsordnung - RVO)
- at the workplace level, legislative measures are also taken by virtue of the involvement of works councils in health and safety at work; the legal basis is provided by the ASiG and the BetrVG.

Issues and workers covered

The law on work protection consists of two bodies of legislation - social protection and technical protection. Social protection at the workplace itself involves two main issues: working hours ('working time') protection and the protection of particular groups of people. Technical protection is provided by both governmental and autonomous legislation; the latter category consists essentially of Accident Prevention Regulations (Unfallverhütungsvorschriften – UVV). The regulatory purpose of technical (safety) law is to protect against the risks and hazards which may arise from production equipment and facilities, from the goods produced and from other technical sources of danger. It includes measures to prevent accidents at work, to minimise occupational illnesses or other medical conditions influenced by working conditions, as well as to legislate for workplace hygiene and other medical protective measures.

The implementation of European Directives

The effect of the Framework Directive (89/391/EEC) is to compel the federal government to redraft a wide variety of regulations on health and safety at work as part of federal law. According to provisional measures drawn up by the ministry responsible (BMA), the EU requirements will be met by updating the ASiG and making changes to the RVO. As required by the Directive, the intention is to add details of 'the fundamental duties of employers and employees with regard to health and safety at work' to the ASiG.

Occupational health and safety structures

The MIAs (Berufsgenossenschaften) and the 'trade supervisory authorities' – which historically have always had the function of a factories inspectorate – are the executive agencies of what is effectively a dual system of occupational health and safety in Germany. The trade supervisory offices are answerable to the provincial ministries of labour and their responsibilities are divided on a regional basis. Any single trade supervisory office thus monitors all businesses within its area, regardless of the industries in which they operate.

The MIAs, or accident insurance underwriters, are divided into three groups: there are 20 agricultural and 36 general trade MIAs, plus 51 internal accident insurance agencies for the public services, operating under broadly similar laws to those of the other two groups. These are all self-administered bodies, with management boards composed of employer and employee representatives in equal proportion. The general trade MIAs are organised to cater for specific industries or branches of the economy. Their tasks are to provide insurance against accidents at work or on the way to or from the workplace, to issue accident prevention regulations, to advise firms on the implementation of such regulations and to supervise their implementation.

The third fundamental pillar of the German system of 'work protection' alongside the MIAs and the trade supervisory offices is the organisation of health and safety within firms themselves. Employers are obliged by the ASiG to establish health and safety systems with an adequate level of staff qualification. This normally involves the deployment of safety officers and company doctors to investigate potential and actual causes of accidents or work-related illnesses, and to stipulate what protective measures should be taken.

Other bodies involved in health and safety at work include the regionally organised Technical Control Boards (Technische Überwachungsvereine - TÜV) which operate under private law and are vested with quasi-sovereign powers associated with the compulsory inspection of plant and equipment. The standards' association, Deutsches Institut für Normung (DIN), is a private incorporated body which has acted as the central standards-issuing organisation in Germany since 1975.

Control and inspection

Mission

The control and inspection bodies perform a dual function. The trade supervisory offices and the MIAs not only inspect and monitor the observance of occupational health and safety regulations, but also encourage the firms they deal with to implement the regulations by providing them with expert advice.

Field of competence

Under the terms of Section 546 of the RVO, the MIAs are obliged to apply all suitable means to ensure that accidents at the workplace are prevented. All issues relating to that task fall within the associations' field of competence. The field of competence of the trade supervisory offices takes in all regulations on health and safety at work, and has also been expanded in recent years to include the inspection, monitoring and implementation of environmental protection legislation.

Inspection

In 1992, the inspectors of the MIAs made a total of 596,673 inspections in 355,578 different companies. The number of officials in the MIAs actively carrying out inspections was 2,056 in 1992, covering a total of 2,537,550 participating firms with an input of 28,484,749 person-years, giving a ratio of approximately 13,854 person-years per inspector. Figures for the trade supervisory offices are slightly better: a ratio of 6,596 employees per inspector can be derived from figures available for 1992.

The majority of the inspecting staff of the MIAs, the technical supervisory officers, are engineers. Their training consists chiefly of supervisory and advisory activities in firms, but it also includes courses on technical, legal and occupational health matters, further courses on communication and teaching techniques, and, importantly, an information-gathering period spent in the administrative departments of the MIAs. The educational requirements for the trade supervisory authorities vary according to the grade sought; for the highest, executive grade, a university degree in a technical subject is required. During the two-year preparatory service that follows acceptance into the grade, training is provided in law, occupational health and hygiene.

Number of sanctions per company

In 1992, statistics show a rate of 0.76 sanctions per company for trade supervisory authorities and 0.33 sanctions per company for the MIAs.

Occupational health services

The organisation of company health services has been covered by the ASiG since 1st December 1974. Company doctors examine new employees before they commence work. They also investigate the threshold at which medical conditions can be triggered off by hazardous chemical, physical or biological influences. If the doctor finds that an examinee is being placed at a medical risk by his/her working conditions, he/she must see to it that the matter is remedied. Company doctors also provide advice to the employer, to the works or staff council and to the workforce in general on all matters of preventive health care. They need to have studied for six years and completed a medical degree. They can then go on to a four-year postgraduate training programme to obtain the title of Doctor for Occupational Health (Arzt für Arbeitsmedizin). The supplementary training for the company doctor's qualification lasts for two years.

The tasks of the safety officer are to advise on:

- the operation of plant and machinery
- how to use tools and equipment, materials and protective clothing or devices
- how to organise production processes to take best account of health aspects.

Safety officers also inspect plant, machinery and equipment to ensure that they are safe. They make observations and surveys of the production processes within the company, evaluate those observations and report potential health hazards to their superiors or to the health and safety committee, and they train employees in the safety approach to the technical facilities with which they work.

Financing

All costs relating to health and safety within the firm must be borne directly by the employer.

Information and assistance

Several of the MIAs issue periodicals. The MIA trade federation (HVBG) publishes a scientific periodical, DBG, for engineers and scientists in the field of occupational health and safety. In addition to a variety of publications in book form, the federation also distributes the periodical *Arbeit und Gesundheit* to technical schools and to companies; it is published monthly, with a circulation of 48,000. The engineering and electrical engineering federations, VDI and VDE, are both professional corporations with regulatory powers in their particular fields. They prepare technical standards. The VDI publishes approximately 1,500 guidelines in 25 handbooks, each with a circulation in the range 400–2,000. Booklets containing a total of 200,000 standards and guidelines are available from booksellers. These are aimed chiefly at the manufacturers of technical equipment, who are required to observe those standards in design and construction. Similarly, the VDE (for the electrical engineering industry) publishes the *VDE Compendium of Regulations* (*VDE-Vorschriftenwerk*), plus an annual report of the work of the VDE and the German Standards Association (DIN).

Training

In 1992, the MIAs staged a total of 15,395 courses with 341,995 participants. A substantial proportion of the basic, advanced and refresher training schemes for people occupying specialist positions in the occupational health and safety system are devised and provided by the MIAs, although several federal agencies are also involved in such training

Economic incentives

Trade Supervisory Authority

Breaches of the working-time regulations are normally subject to a fine of up to 5,201 ECU, violations of the technical health and safety rules up to 40,816 ECU and environmental infringements up to 52,016 ECU.

MIAs

The MIAs are empowered to initiate summary proceedings if accident prevention regulations have been breached. Each infringement can be fined up to 10,403 ECU.

Range of variation of risk-related premiums

Firms have an economic incentive to improve prevention since the MIAs set performance-related bonuses and surcharges. The total value of the bonuses granted in 1992 was 109.9 million ECU (1.35% of the regular mutual levy), and that of the surcharges imposed was 350.86 million ECU (4.32% of the levy). The total mutual levy raised by the associations was 8.119 billion ECU.

Overall yearly budget for subsidies and grants

An estimate is that the maximum public funding available for improvements in occupational health and safety is in the order of 34–42 million ECU. The bulk of this money is made available by the Federal Ministry of Research and Technology (Bundesministerium für Forschung und Technologie – BMFT) under its 'Work and Technology' scheme ('Arbeit und

Technik' - AuT).

Research

Among the most important bodies researching into industrial safety and occupational health in Germany are the Federal Agency for Industrial Safety and Accident Research (BAU), the Federal Agency for Occupational Health (BAfAM) and the Dortmund Social Research Unit (SfS).

Occupational health insurance

The mission of statutory accident insurance (SAI) (gesetzliche Unfallversicherung) entails:

- prevention of occupational illnesses and workplace accidents
- provision of first aid in the event of workplace accidents
- restoring injured persons' ability to work and promoting their return to their occupations
- paying out compensation.

The fundamental principle followed in the exercise of these tasks is always: prevention before rehabilitation and before compensation.

The SAI scheme is funded entirely by the contributions paid by firms, as they themselves bear the responsibility for accident and health hazards associated with their activities. The Accident Insurance Act (Unfallversicherungsgesetz - UVG) stipulates that all companies or business proprietors must join an MIA, which covers the costs of workplace accidents regardless of who is at fault. The financial outgoings of the MIAs are met on a pro rata basis by all the associations' members. This cost the firms a total of 7.831 billion ECU in MIA contribution in 1992 (1.43 ECU per 100 ECU of compensation). In 1992, the total compensation awarded by the MIAs was over 6.2 billion ECU and they spent 458 million ECU on accident prevention, occupational health services and first aid in 1992. The UVG Act empowers the MIAs to issue their own accident prevention regulations and to have their own technical inspectors monitor the observance of those regulations.

The accident insurance agencies in Germany are not private-sector underwriters, but have the legal status of public-law corporations under the supervision of the Federal Ministry of Labour. The legal status of SAI within the overall health and safety system is peculiar to Germany. In 1991, a total of 50.539 million people were covered by SAI, 38.459 million of whom had that cover provided by the MIAs.

Monitoring

All fatal accidents and any which render an employee unfit for work for more than three days must be reported according to statutory obligations (RVO, Section 1552). This accident reporting procedure was introduced for all statutory accident insurance agencies by an administrative regulation dated 1st January 1974. The organisation of accident reporting procedures within individual workplaces is regulated by Sections 9 and 10 of the ASiG. These require all accidents to be reported without delay to the personnel department within the firm involved in the work protection system.

The most important agencies when it comes to reporting on accidents and health in the occupational field are the National Association of Works Health Insurance Funds (Bundesverband der Betriebskrankenkassen), the MIAs, the Federal Ministry of Labour and Social Affairs (BMA), the Federal Office of Statistics (Statistisches Bundesamt) and health reporting systems now being established in the individual provinces.

Depending on their own fields of emphasis, the BMA, the works health insurance funds and the HVBG mainly report on accidents at the workplace or on the way to/from work, occupational diseases, pensions, expenditure and health and safety regulations. The regional health reports also cover health hazards, the state of public health, health protection measures and rehabilitation.

3 Health and safety outputs

Occupational accidents

The total number of accidents reported in 1992 (by public service internal insurance agencies, agricultural mutual indemnity associations, and trade MIAs) was 2,069,422 of which 1,622,732 occurred within the sphere of the trade MIAs. The frequency of reported occupational accidents was 55 per 1,000 person-years.

The incidence of occupational accidents is mainly concentrated in those industries involving heavy-duty work processes. In relative terms, the highest accident frequencies (i.e. mandatorily reported accidents per 1,000 person-years) occur in the construction industry and trades (124.0), in timber processing (121.4), in the rock, stone and related mineral products industry (84.6), the metal industry (80.3) and the mining industry (76.9).

The most important trend as regards accidents, which persisted into the 1980s, was a continual decline in fatal accident risks, both in absolute terms and in relative terms (per 1,000 person-years). The number of fatal occupational accidents compensated decreased from 3,021 in 1960 to 1,086 in 1990.

Occupational diseases

In 1992, 85,721 separate reports were made of suspected occupational illness, which break down as follows:

- skin afflictions - 24,056
- impaired hearing due to noise - 12,243
- allergic respiratory complaints - 5,903
- asbestosis - 2,954
- silicosis - 2,924.

The same classes of illness occupy the top positions in the statistics for recognised cases of occupational illness. There are 59 recognised occupational diseases in Germany. Proof of a causal link between occupation and disease is normally required for recognition. In 1992, the MIAs recognised a total of 11,308 cases of occupational illness in 1992, which included 3,780 cases of impaired hearing (33.43%), 2,716 of skin afflictions (24.02%), 1,055 of allergic respiratory complaints (9.33%), 646 of asbestosis (5.71%) and 558 of silicosis (4.93%). The decline in the number of cases of silicosis, which had been the classic example of an occupational disease right up to the 1960s, is largely attributable to the drastic fall in the number of people employed in the coal, iron and steel industries. Even with the improved preventive measures of today, however, the risk of silicosis remains high among miners. Hearing impairment, a statistically very prominent occupational hazard in the 1970s, now appears to be levelling off in its incidence at a relatively high level. A very noticeable development in recent years has been the pronounced increase in allergic skin complaints and obstructive respiratory disorders. Asbestos-induced occupational diseases are also growing increasingly significant.

The greatest problem with regards recognising and combating occupational disease is the long exposure period of anything up to 30 years that is typical of the most significant illnesses. In other words, the occupational diseases being identified today are largely a reflection of past working conditions. The implication for health protection and safety measures at the workplace is that preventive measures can only be geared to occupational illness statistics to a very limited extent.

Absenteeism

Sickness leave

The number of cases of unfitness for work showed a marked fall for the first time in 1992, at 148.9 cases per 100 insurance fund members. The number of working days lost per 100 members in former West Germany was 2,351; again, well below the previous years. However, the average duration of unfitness for work remained the same, at 15.8 days.

Invalidity

The trade MIAs made a total pay-out of 4.03 billion ECU in pensions in 1992 to injured or sick employees and to the families of those who had died; 70% of that sum applied to injured or sick employees who were still alive. Allowances paid to surviving dependants totalled 11.9 million ECU and compensation payments were 76.9 million ECU.

The economic cost of occupational accidents and diseases and absenteeism
Absenteeism due to illness alone has been estimated by BAU to cost the economy an annual amount running into three figures of billions. Skeletal disorders alone are estimated to have an aggregate economic cost of approximately 11 billion ECU. The total loss of resources due to occupational accidents (also including those on the way to or from work) is estimated at over 18 million ECU.

4 Assessment of occupational health and safety policies

The strengths of the existing system of work protection were pointed out by all the experts interviewed. These, they said, are:

- the high level of qualification of the officials dealing with occupational health and safety
- the standardised mission of the statutory accident insurance bodies
- high safety levels in technical industrial standards.

The chief deficiencies in the existing system were found to be:

- the coverage of SMEs, particularly in the poor conditions for developing systematic epidemiology
- the system being chiefly geared to typical accident hazards.

There was also a considerable degree of consensus that more preventive measures and monitoring programmes need to be established, especially in fields with high inherent risks. The need for action is particularly urgent in the field of work-generated illnesses. The objective is to establish a system of comprehensive, preventive employee-health protection.

Although the MIAs have not observed any slackening of safety standards as a result of the recession, pressure to save has had a marked impact on the trade supervisory authorities and on employers hit by the recession. Also needing to be taken into account is the fact that employee protection takes second place compared to employment security and that economic competition is frequently used as an argument for introducing more flexible working times. As well as the classic factors impairing employee health, psycho-social stress, hazardous substances, surveillance of construction sites and ways of achieving greater involvement of employees themselves in the protection of their health and safety at the workplace were cited as likely to be amongst the more significant problems in the future.

The EU's influence on work protection is mainly seen in a positive light in that it will improve the scope of legislation and also the recording of ill-effects on employee health. However, the implementation of EU Directives is also associated with a weakening of national regulatory bodies and it is feared that plant and machinery may be granted approval which would not have passed currently applicable national requirements.

Conclusions

A consensus is now forming in Germany, partly spurred by EU Directives, that combating psycho-social risks (e.g. lack of decision-latitude, social isolation) and promoting health potential (e.g. health information, social support) will gain in significance, and will play a substantial part in future preventive health and safety, alongside the traditional tasks of technical and social safety at work. Consensus is also emerging to the effect that systematic monitoring of the health situation of employees will become an indispensable instrument of health protection at the workplace.

Another view receiving growing attention is that health promotion (rather than merely combating health risks) should become a central objective of health and safety at work. Health promotion at the workplace should be regarded as one element of a company-wide policy on health. In other words, health promotion should become the core feature of a comprehensive development of human resources. With regard to the changes in the pattern of illnesses and also to the changing structure of the workforce, many experts believe that a healthy design of tasks and organisational structures ought to be given priority over behaviour modification. Greater attention should be paid to age- and gender-specific health problems, as well as to health promotion in the service sector, especially in public services. These new tasks can only be mastered with the aid of comprehenshive training programmes for the people responsible for workplace safety and health, and with intensified, inter-disciplinary co-operation amongst them. Quality assurance and the proper evaluation of programmes and measures taken need to become routine in occupational health promotion. On matters of work and health, employees themselves are experts on their own situation. As a 'bottom-up' strategy for modern health protection, health circles have proved their worth in several different organisations, and deserve to be more widespread.

The organisation of work protection in Germany is regarded as rather cumbersome. There is some criticism of the 'dual system', due to the continuing problems of having governmental bodies and para-fiscal bodies working side-by-side. Since 1989, the statutory health insurance funds have also come onto the scene with the additional task of health promotion at the workplace - a task they are now taking increasingly seriously. The insurance funds are about to make a considerable contribution to the modernisation of health care in the workplace; on the other hand, they are also helping to make the overall picture even more complex, with concomitant problems of co-ordination and co-operation.

Just recently, a move to modernise Germany's health and safety at work law was thwarted by resistance from the liberal FDP parliamentary party. Having so far failed to adapt its occupational health and safety legislation to comply with the EU's Framework Directive, the German federal Government is running the risk of having a case brought against it at the European Court of Justice.

1 The context of national occupational health and safety policies in Greece

Economic structure

The economy of Greece is dependent on the service sector which represented 56.5% of the GDP and employed 50.3% of the total labour force in 1991. Tourism is one of the most important branches of the Greek economy. The industrial sector of the country is not much developed; it represents a little more than 25% of the GDP and employs a similar percentage (27.5%) of the workforce. Its share in the Greek economy has, however, remained stable in the last five years whereas agriculture has diminished over that period. Agriculture contributes an approximate 15% to the country's GDP and employs between a fifth and a quarter of the labour force.

The Greek economy is characterised by a large public sector. Public expenditure represented 47% of the GDP in 1989, a percentage that has been steadily increasing since 1958. A serious effort is being made to lower this with some success in recent years; there was, for example, a 2.3% fall in the number of state employees in 1991 and an 8% fall in 1992.

Small and medium-sized enterprises (SMEs) overwhelmingly dominate the Greek market. In 1988, 96.4% of all enterprises operating in Greece employed under 10 workers, amounting to a total of 56.5% of the labour force, with 74.2% being employed in enterprises with 50 or less employees.

The labour market

The population of Greece is ageing and the birth-rate has been consistently low too. According to official statistics, unemployment has remained below 10%; however, the 1990s have seen an increase, with the rate reaching 9.3% in 1992, having been 7.7% in 1988.

3.5% of the labour force is in precarious employment - a percentage which has remained stable in the last few years. According to official statistics, the number of people in multiple jobs is decreasing (5% of the workforce in 1988 to 3.8% in 1993), while part-time employment is also decreasing. Perhaps the most outstanding characteristic of the Greek labour market is the high percentage of self-employed people (more than 25% of the workforce). This is a trend which has not changed since 1988.

The employment of young people (under 20 years old) is approximately 3.5%, but it is decreasing. Older labour has remained stable at approximately 15.5% for workers 55 to 69 years of age, and 34% for workers between 45 and 69.

Women represent approximately 37% of the total labour force, most of whom (56.3% of the female labour force) work in services. Women's unemployment is under-reported, but even so, from 12.3% in 1989 it reached 14.2% in 1992 and is increasing; it is 1.5 times higher than male unemployment.

Migrant labour has increased considerably in recent years, especially in the informal economy. Although there are approximatly 35,000 legal working migrants, the number of illegal migrant workers is much higher, estimated at approximately 500,000. Most of them work in construction, parts of the manufacturing and service sectors, agriculture and as domestic labour. Illegal migrant workers come to Greece primarily from Eastern Europe countries and poor countries of the south (e.g. Pakistan, India, the Philippines, etc.).

Industrial relations

Collective bargaining is mainly determined by legislation. It has been enriched and further developed with the passing of Law 1892/1990. Since then, industry-level and company-level collective agreements have increased in number. Furthermore, employers' and employees' co-operation on occupational health and safety (OHS) and other work-related issues has expanded. A new co-operative attitude is becoming more commonplace. Union membership is limited and declining in both the public and the private sectors (from 23% in 1989 to 19.3% in 1992). Employers are organised in several federations.

Workers' councils and OHS committees or OHS representatives are established by law at enterprises employing more than 20 people. In the context of the Greek market, these institutions only play a marginal role since the number of SMEs is very high. However, their potential contribution to OHS prevention is noteworthy.

The mechanisms and institutions for OHS prevention exist in Greece, but co-operative relations and the social dialogue which has started between employers and employees must be considerably promoted if they are to be effective. Appropriate structures, infrastructures, attitudes, information dissemination systems for the promotion of OHS issues require cultivation; appropriate OHS training for both employers and employees must be carefully and systematically organised if existing institutions are to become fully effective.

Values and philosophy in relations to health and safety

OHS issues have not been a high priority for the state, employers or employees; other issues - different for each participant - have taken precedence. Where OHS is a low priority in industrial relations, the implementation of OHS legislation is not vigorous. Despite legal requirements, only a small number of those enterprises obliged to appoint safety officers and occupational doctors to undertake serious occupational risk prevention activities actually have them. Even fewer have appropriately staffed and well equipped OHS units. Furthermore, workers' participation on OHS committees is limited, partly because of the management style of enterprises in Greece - which is characterised by low-level industrial democracy - and partly because of workers' concern for 'bread and butter' issues.

However, in 1988, an important change occurred in industrial relations which until then had been consistently confrontational with strong state intervention. That year marked the beginning of dialogue between employers and employees with the aim of finding consensus solutions to OHS problems. A Joint Experts Committee was set up to draft a document which identified the basic problems in OHS and made proposals for the future development of health and safety in Greece.

2 Occupational health and safety policies and structures

Legislation

Until 1985, OHS legislation in Greece was antiquated and inadequate for the needs of the times. Law 1568 passed in that year provides the general philosophy which has guided OHS matters since then. Under this law, OHS prevention is the legal responsibility of each individual employer, especially of employers in enterprises where there are more than 150 employees. These large enterprises are legally obliged to have a safety officer and an occupational doctor. Furthermore, OHS is the inalienable right of workers: they have the right to have OHS committees as well as workers' councils. Inspection of occupational health and safety at all workplaces is the responsibility of the Ministry of Labour.

The OHS participation system, determined by Law 1568/1985, provides for organisation at three levels:

- workers' participation in OHS committees or workers' councils and regular (every 3 months) OHS consultation at the level of the individual enterprise
- a prefecture OHS committee in which the central government (the Technical Inspectorate of the Prefecture), the labour centre and the Technical Chamber of Greece, participate at the regional-prefecture level
- the OHS Council, at which all social partners and professional organisations are represented, at national level.

All OHS participation institutions are advisory at all levels.

Law 1568/1985 spells out the duties and responsibilities of employers, workers and the manufactures or importers of machinery and chemicals. It contains detailed provisions on the protection of workers from physical, chemical and biological agents and it established workers' right to information about occupational hazards in the workplace.

Other existing OHS legislation covers a variety of different topics. Some decrees specify in detail the obligations of employers for the protection of workers' health and safety in certain trades and professions and under specific working conditions. Other decrees regulate the work conditions in various sectors of industry. Yet other laws focus on specific hazards in the workplace and identify health and safety requirements. OHS regulations and laws are fragmented and codification is required to improve their accessibility and hence usefulness.

The European dimension

All European Directives for health and safety which were passed in the period 1986 to 1989 have been transcribed to Greek legislation. The obligation to implement the Framework Directive 89/391/EEC and its daughter Directives has generated much discussion amongst the social partners and professionals. The transcription process is expected to be concluded by the end of 1994. EU Directives on OHS issues have unquestionably had a positive influence on Greek legislation, but the influence of EU policies on attitudes and practice concerning OHS issues is limited and the implementation of OHS laws is not vigorous.

Occupational health and safety structures

The Ministry of Labour is the main controlling body for the work environment in all work activities, public and private, but other ministries have some responsibilities, for example in the mining and quarrying sector and on board ships, whether at sea or anchored in ports. Recent developments, however, have transferred some of the Ministry's OHS responsibilities to the prefectures.

Control and inspection

Under new developments in early 1995, the administrative structures which share responsibility for OHS prevention at the level of the Ministry are:

- the OHS Council
- the Centre of Health and Safety at Work (KYAE)
- the Directorate of Working Conditions.

At the level of the prefecture they are:

- the units of technical and health inspection (approximately 50 units overall)

- the centres for the prevention of occupational hazards (KEPEK) (only five, located at the Prefectures of East Attica, West Attica, Piraeus, Athens and Thessalonnika).

The mission of the Ministry's OHS-related directorates includes OHS policy-making, research and the measurement of physical and chemical factors in the work environment. The mission of the prefectures' OHS-related directorates or units is primarily technical, including health inspection of workplaces, prevention of occupational hazards and implementation of OHS legislation. Their fields of competence cover all work environments, except mining, quarrying and on board ships.

According to official data, an approximate 2% of enterprises are inspected annually and approximately 1% are re-inspected. SMEs are insufficiently inspected at present. No data exists on the inspection by employment sector.

The training of inspectors

The Ministry of Labour employed 166 technical inspectors in the central and regional services in the period 1989 to 1993. Of these, 77 had university degrees, 51 held degrees from higher technical schools, and 38 had high school qualifications. Their training on health and safety consisted of occasional seminars organised by the Ministry of Labour in collaboration with the National Centre for Public Administration. There are real training needs for technical inspectors. They have insufficient skill in measurement-taking in the workplace and in conducting safety audits. They have no training at all for making technical suggestions or in communication skills; these skills depend entirely on the individual inspector's initiative and personal aptitude.

There are 174 vacant posts for technical inspectors. In manufacturing and construction where there are 750,000 employees, there is currently one inspector for every 4,518 workers. Of the workplaces inspected between 1989 and 1993, sanctions were imposed on between 2 and 4%.

Occupational health services

At governmental level, only the Technical and Health Inspectorates and the Hellenic Social Security Institution (IKA) provide services on OHS-related issues. Technical and Health Inspectorates advise on and control OHS implementation. IKA monitors specific health hazards - lead, respiratory problems, heavy metal dusts and so on - at the workplace on request.

At the level of individual enterprises, although legislation (Law 1568/1985) requires the employment of safety officers and occupational doctors in enterprises with more than 150 employees, only a small number of these enterprises have prioritised occupational risk prevention highly enough to produce noteworthy results. This means that only 200,000 out of the 706,308 employees (28% of the workforce) in the manufacturing sector are covered by OHS services. The percentage of the workforce covered in services and commerce is even smaller.

The Institute of Research of Chest Diseases and Occupational Medicine, as the first of a series of advisory centres on health and safety at work, is expected to play a role in the promotion of OHS services, especially to SMEs at regional level.

Occupational doctors carry out preventive medical examinations, give written advice to employers and to workers' OHS committees, and provide first aid in the event of an accident or an emergency. However, only 25 of the doctors who are employed as occupational physicians are actually trained in occupational medicine. The rest (nearly 400) are specialists in one of the following: internal medicine, thoracic medicine, cardiology, surgery or they have no specialism. Approximately 100 have attended special, three-month long training seminars in OHS.

Safety officers are legally required to have a university degree. Although in practice they meet the formal requirements of the law, they often lack experience and specific training in OHS, in addition to lacking enough time to carry out their duties, since employers often recruit safety officers from amongst existing employees.

At national and prefecture level, OHS institutions are financed from within the framework of the government budget. Occupational doctors and safety officers employed at individual enterprises are paid by the enterprises themselves.

No direct or systematic link has been established between occupational doctors in enterprises or between the Technical Inspectorates and the National Health System. Only the OHS services that IKA offers are encompassed in its general health package.

Information and assistance

There is no single information centre or library that covers the whole range of topics included under OHS. Information on specific subjects may be acquired through one of the following information centres:

- the National Documentation Centre, based at the National Research Foundation (electronically connected with 48 Greek libraries and with 800 international data banks)
- the Department of Documentation and Information of the Library of the Technical Chamber of Greece
- the Centre for Greek Medical Information, Terminology and Documentation (IATROTEK)
- Euro-Info Centres of CEC-Taskforce SMEs.

This lack of an integrated OHS information infrastructure in Greece will be resolved when a nationwide OHS information system which ELINYAE plans to create is operational.

According to OHS legislation, the employer is legally required to give information on OHS hazards at the workplace to the Technical and Health Inspectorates and to employees whenever it is requested. When the 89/391/EEC Framework Directive becomes incorporated into Greek legislation, this provision will become more vigorous since employers will be required to provide such information irrespective of whether it has been requested.

OHS information dissemination initiatives have been undertaken by a variety of interested parties including the Ministry of Labour, scientific associations, employers' and employees' federations, the various trade chambers as well as individuals.

Training

Training courses which include OHS subjects are either organised by private enterprises for their employees, or organised in the public sector for public servants. Official data on OHS training is not available, but it is estimated that only 15% of the total duration of employment training programmes is related to OHS. After the implementation of Law 1568/1985, training was provided for professionals in preventive services and for safety committee members by several institutions, including the Ministry of Labour, the social partners, professionals and management. The need for the preparation of OHS experts at all levels of the educational system has still not been met and there have been only occasional efforts to transmit OHS knowledge at the level of secondary education.

Economic incentives

The economic incentives that support OHS prevention in Greece are penalty fines, insurance premiums and subsidies or grants.

The annual number of penalties imposed on enterprises in the period 1988 to 1992 fluctuated from 61 to 185. The average value of these penalties is estimated to be 700 ECU each.

Insurance premiums for OHS are included in social security premiums. Special premiums are paid for employees working in heavy and unhealthy occupations. There is no formal mechanism, however, to link the insurance premiums paid by enterprises with their performance in OHS prevention; although in the case of private insurance companies, OHS premiums increase after the occurrence of industrial accidents.

The state gives subsides or grants to enterprises that undertake investments for the improvement of the well-being of workers. However, there are no data on the level of those subsidies.

Research

OHS-related research in Greece has covered a wide spectrum of issues. Most of this research has been funded by the European Union, the Greek Government, private enterprises and other organisations; a few were self-financed. No data are available on the budget for OHS research projects. Engineers, chemists and doctors have carried out epidemiological studies, measurements of risk factors at specific workplaces, studies of occupational safety, studies on mechanisms and/or methods, policy analyses, general studies on OHS and other kinds or research. The results of such studies are disseminated through symposia, books, leaflets, seminars, etc.

Occupational health insurance

OHS insurance is mainly incorporated into the framework of the general social security scheme. Private insurance companies have entered the market in the last ten years or so; they provide some incentives for OHS prevention on a case-by-case basis. However, only IKA has a structured unit - the Centre for Occupational Medicine - which is well equipped and staffed and able to play an important role in OHS prevention.

Monitoring

The national monitoring system consists of the Technical and Health Inspectorates at the level of government, and occupational doctors and safety officers at individual enterprises. The systematic monitoring of OHS issues, however, is provided only by the Directorate of Working Conditions of the Ministry of Labour and by IKA's Statistics Department. Both of these administrative bodies have electronic data-bases with information on occupational accidents and the condition under which they occurred. Occupational monitoring services are occasionally offered by other departments or units of the Ministry of Labour, the Ministry of Health, universities and private OHS consulting firms.

Emerging issues

The main OHS issues and needs are:

1. The staff of the Ministry of Labour OHS services must increase in number. They must be better equipped, receive improved and continuous training and develop a four- or five-year management plan with specific objectives. An information network should be established for the technical inspectors and the personnel of the Ministry of Labour OHS departments.

2. The OHS units of most enterprises are not well staffed and equipped.

3. The coverage of SMEs with OHS services is non-existent. For this reason the initiation that is underway of the advisory OHS centres for groups of SMEs is of great importance.

4. The link between occupational health services and the national health system must become substantial and effective.

5. The national health insurance system must become prevention-oriented and provide workers with preventive and health promotion services.

6. The financial incentives for enterprises to take and maintain measures for health and safety must be clear and effective. Equally, fines and other sanctions for those who ignore OHS measures must be meaningful.

7. Greek universities must start educating safety officers and occupational doctors as well as the other professionals who are indispensable for a multidisciplinary approach.

8. OHS courses must be introduced at all educational levels. Those OHS training programmes that are targeted towards employers, employees and specialists must have common aims and principles of substance.

9 An effective information dissemination system should be organised.

10 Research on OHS must be co-ordinated.

11 The result of OHS monitoring must be disseminated to all interested parties.

12 The social partners should upgrade OHS on their agendas if they want to avoid the deterioration of working conditions at a time of economic crisis and unemployment. Furthermore, their participation in social dialogue on OHS should be continued and expanded.

13 OHS issues should be examined as a constant element of industrial relations.

3 Health and safety outputs

Occupational accidents

According to the statistics that IKA publishes annually, approximately 32,000 people were injured and 750,000 work days were lost in 1988. Since then, occupational accidents have decreased steadily. The incidence of fatal accidents has fluctuated in the same time period with a general increase during the the last 5 years.

OCCUPATIONAL ACCIDENTS

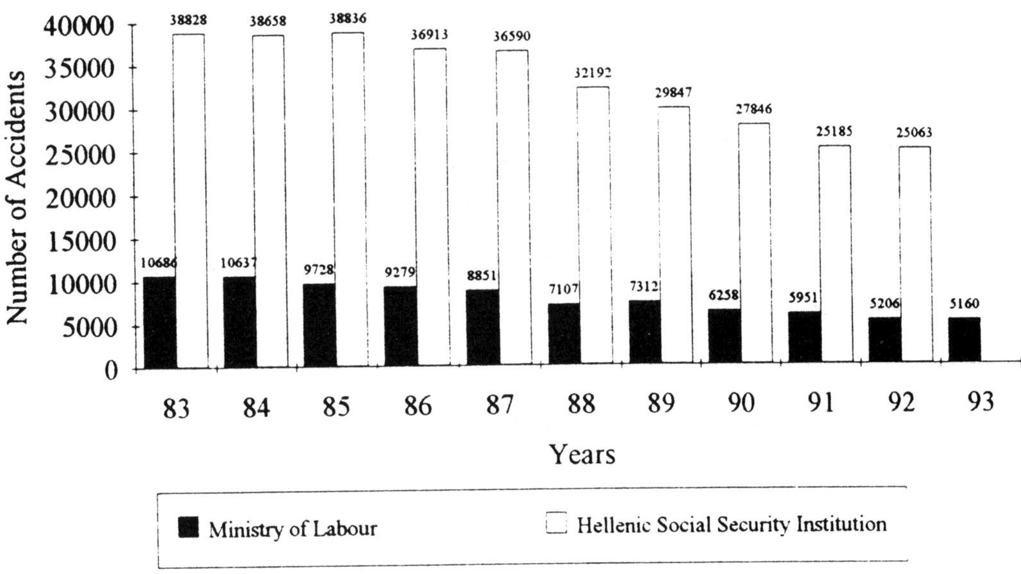

Source: Ministry of Labour (1993), Hellenic Social Security Institution (1992)

Figure 1: Occupational accidents

Between 1988 and 1992, most of the reported accidents occurred in manufacturing (approximately half) and in construction (approximately a quarter). According to 1992 data the most dangerous work environments are mining (294 occupational accidents (OA) per 10,000 employees), construction (270 OA per 10,000 employees), manufacturing (180 OA per 10,000 employees) and electricity (104 OA per 10,000 employees). The most common injuries were external injuries, muscle strains and simple fractures; while arms and legs were the most frequently injured parts of the body.

The direct cost of occupational accidents is registered by the Statistics Department of IKA. There is a considerable need in Greece for the systematic registration of both occupational accidents and diseases, as well as the maintenance of the corresponding statistics.

Occupational diseases

In Greece a disease is considered an occupational disease if it results from undertaking an occupation, but in practice the incidence of occupational diseases is heavily under-estimated because many cases are classified as non-occupational. The IKA Centre of Occupational Medicine has made a contribution towards better diagnosis. The commonest occupational diseases in Greece is dermatitis followed by lead poisoning and pneumoconiosis. Incidence rates cannot be discussed, however, since no reliable data on occupational diseases exist.

Absenteeism

Absenteeism from both general and occupational causes decreased in the years 1990 to 1992, after an increase between 1988 and 1990. The cost of absenteeism caused by occupational accidents is considerable, for example:

1988 8,582,729 ECUs
1990 9,687,380 ECUs
1992 8,515,677 ECUs.

4 Assessment of occupational health and safety policies

The response of the representatives of the interested parties interviewed for this report highlighted the strengths and weaknesses of the OHS system in Greece. They concurred on the following main concerns:

- the need for better staffed, more fully equipped and better administered Technical and Health Inspectorates
- the need for the promotion of the social dialogue and ELINYAE
- the need for the better OHS training of employers and employees
- the need for a good information dissemination system
- the need to improve the severely inadequate OHS services for SMEs.

On the other hand, they applauded the positive developments and strengths of the Greek OHS system, such as:

- the Technical and Health Inspectorates' coverage of the whole country
- the recently begun social dialogue
- the establishment of ELINYAE
- the establishment of OHS committees and workers' councils at plant level.

The main OHS priorities highlighted by the respondents reflect those identified earlier as emerging issues and include the following:

- the promotion of OHS structures for SMEs
- better and more effective inspection and implementation of the law
- adequate OHS training and education for all relevant parties
- the raising of employers' and employees' awareness of the importance of OHS prevention.

The EU's impact on OHS policies in Greece is generally believed to be positive. As the economic crisis and unemployment influence working conditions and OHS adversely, however, those of the EU's economic and social policies that require industrial restructuring and/or further threaten employment, might have a negative impact OHS prevention.

1 The context of national occupational health and safety policies in Ireland

Economic structure

Ireland has an open economy which means that its economic and employment performance is related to international trends; the close relationship between Ireland and the UK labour market is one example of this. For a small country like Ireland, adaptability to the external environment is vital for accelerating employment growth. Characterising the labour market in the 1990s are high levels of unemployment and the prospect of large inflows of school leavers and married women seeking to return to work.

Significant changes of economic and social importance have occurred in Ireland during the past five years. Analysis by economic sector shows that agriculture has declined and it is forecasted to fall further into 1996. Employment in industry has shown a steady increase, particularly in the high-tech area of manufacturing. The services sector is steadily increasing and this trend is forecasted to continue up to 1996. Small and medium-sized businesses (SMEs) have been neglected during the 1970s and 1980s but the Government is now recognising the need to support them. Some 24% of the workforce are in the public sector which has shown rapid growth over the past five years.

The labour market

While the overall participation of women in the labour force in Ireland is relatively low compared with other EU countries, it has grown steadily and reached 32.8% in 1992, with married women constituting an increasing proportion. Ireland has a long history of unemployment which rose to 16.3% of the workforce in 1993. Part-time work has also been on the increase in recent years.

Industrial relations

The conduct of collective bargaining can vary considerably between organisations but remains essentially voluntary in nature. For unionised organisations, an important influence on the practice of employee relations is the level at which collective bargaining is conducted:

multi-employer or single-employer level. The traditional focus has been on multi-employer bargaining with supplementary bargaining at workplace level. Here, the views of the employer on employee relations issues, particularly pay, are represented through the appropriate employer association.

Single-employer bargaining became established in Ireland in the 1960s. With the advent of highly centralised bargaining arrangements in the 1970s, a clear differentiation was created between issues for negotiation at national level and at establishment level. General pay increases were negotiated at national level, while the emphasis at local level was on dealing with productivity and working conditions.

The national pay rounds have been dominant since the Industrial Relations Act, 1946. Between 1946 and 1981 there were twenty-one wage rounds – thirteen centralised and eight decentralised. National understandings were introduced in 1979 and 1980. These consisted of a tripartite understanding – government, employer and industrial organisations, and the ICTU – relating to economic and social development. The agreement on pay policy was part of the National Understanding.

By the early 1980s, the National Understandings expired and industry-wide bargaining disappeared. More and more companies were successfully pleading inability to pay the full terms of the agreement. It appears that many individual agreements were made at the level of the firm during the course of national agreements.

Another feature of collective bargaining which emerged was third party intervention, i.e. the Labour Court, the conciliation service of the Court, the Rights Commissioners, etc.

By the end of the 1980s, a tripartite agreement (Government, unions, employers) called the Programme for National Recovery (PNR) (1987), was agreed. This provided for centralised pay guidelines while allowing management and employees to negotiate on other issues at establishment level.

By 1990, PNR was succeeded by the Programme for Economic and Social Progress (PESP) which was designed to provide a strategic framework for the 1990s and was agreed by the Government and the social partners. It maps out a draft agreement on pay and conditions between the Irish Congress of Trade Unions and Employer Organisations for 1991–1993. It also details the need both for a radical development of the health system and for the positive promotion of health.

Workers' councils

It is argued that the wide scope of collective bargaining in Ireland has probably hindered the progress of representative worker participation. Entry into the European Community was a catalyst to the employee participation debate. This resulted in much discussion and activity throughout the 1970s and 1980s, more clearly manifested in the passing of the Workers' Participation (State Enterprise) Act 1977 which introduced broad-level participation to seven semi-state companies (public sector).

Direct participation by workers either through board-level participation or through workers' councils is a slow development in Ireland. Where workers' councils have been developed, it is recognised that not enough care went into their formation, membership, training and role, and often they have become mere 'talking shops'. Where an enterprise has a number of unions servicing its employees, representatives on the workers' councils tend to be allocated in proportion to the numerical strength of the unions in the enterprise. In 1988, the Workers Participation (State Enterprise) Bill was enacted. This Act has stimulated a renewed interest in worker participation in the public sector. In the private sector, progress on employee participation is slow – whether workers' council, quality circles or other. Some individual companies have taken initiatives, but overall there has been little development since 1987.

Health and safety committees

A major change introduced by the Safety in Industry Act (1980) was the obligation on organisations to have safety committees or representatives. These provisions were subsequently repealed and replaced by new provisions in the Safety, Health and Welfare at Work Act (1989). Despite a campaign by the Industrial Inspectorate and the National Industrial Safety Organisation (NISO), the voluntary system of safety committees encouraged by the 1955 Factories Act never took root. The 1980 Act made mandatory the use of a safety representative or a safety committee. The emphasis was on a consultative process, giving committees an opportunity to assist management in establishing safety policy. Organisations employing more than twenty employees *must* appoint a safety committee comprised of management and employee representatives, whilst smaller organisations *may* have a safety committee or a safety representative. There are no guidelines or legislation on the structure of safety committees.

The main objective of safety committees and representatives is to facilitate workplace co-operation and communication on health, safety and welfare matters, and to agree on the content and application of the organisation's safety statement.

Union membership

There were approximately 477,900 trade union members in Ireland in 1992. This marks an increase of 20,000 over the previous five years, including an increase of 3,000 in 1992. These figures reversed the decline of the recessionary years 1980–1987 when the overall figures fell from a high of 527,200 in 1980 to 457,000 in 1987. Over 55% of employees are members of trade unions, with 48% of total trade union members in the private sector and 52% in the public sector.

There were 70 unions affiliated to the Irish Congress of Trade Unions (ICTU) in 1992, compared to 90 unions in 1970. There are currently 52 unions and the number has been reducing slowly through rationalisation. A major priority for Congress and its affiliated unions is to expand trade union organisation, particularly in those sectors in which membership levels are currently low. In recent years recruitment has been targeted in the engineering sector and the construction industry.

Employer federation membership

IBEC, the Irish Business and Employers Confederation, represents the coming together in 1993 of the country's two major business and employer organisations – the Confederation of Irish Industry and the Federation of Irish Employers. Representing 3,700 firms employing 300,000 people, IBEC is a representative and service organisation. Its mission is to influence the formation of policy at national, European and international levels. IBEC also provides quick response assistance, information, advice and representation for members in industrial relations, pay bargaining, occupational health and safety, economic, legal and commercial issues.

Values and philosophy in relation to health and safety

The values of the social partners, public authorities and other interested groups were reflected in the report of the Commission of Inquiry on Safety, Health and Welfare at Work, 1983. There is general agreement that an integrated programme of advice, information, participation and enforcement needs to be developed for the efficient implementation of the health and safety system in Ireland and that the law should provide a framework to make work safe.

The social partners and public authorities have stressed the importance of training to the improvement of prevention in health and safety, and they emphasise in particular the use of initiatives that are well organised and which help people to learn on a continuing basis. They suggest that training and development will be needed on a scale much greater than has been the case in the past if the objectives of the legislation are to be realised quickly and efficiently. The overall objective of training is to ensure safe working habits and the ability to react preventively both to recognised hazards and to new problems.

Another message which the social partners and public bodies emphasise is that ignoring or flouting safety and health law can lead to prosecution. They feel that responsibility for safety and health at work flows from the highest level of management, through the supervisor and rests on the worker who also has a responsibility. Accident and disease prevention depends on employers managing safety, health and welfare like any other company function, in other words by setting objectives, by planning, organising and controlling to achieve them, by assigning responsibility down the line and by delegating authority to achieve results.

2 Occupational health and safety policies and structures

Legislation

The general thrust of the Irish Safety Health and Welfare at Work Act 1989 follows the approach of the EU. This Act requires management in all organisations to set out in a Safety Statement their own strategy on preventing accidents and ill-health at work. The 1989 Act and the Statutory Instruments which it generated, as well as the strategic work programme of the Health and Safety Authority, are aligned with the EU Directives. The European dimension is viewed as very significant in the Irish context and a catalyst for change.

The underlying philosophy of the Safety, Health and Welfare Act 1989, unlike that of previous legislation, is that it should extend to and protect all workers including the self-employed. Drawing on the mission spelt out in the Barrington Report, the Act is influenced by the philosophy that health and safety are an integral part of management and are best reached by consensus between employer and workers.

The fundamental aim of health and safety legislation is the prevention of accidents and ill-health at the place of work. Employers are responsible for creating and maintaining a safe and healthy workplace and for safeguarding the health and safety of their employees. The law requires that premises, equipment, systems of work and articles for use at work are all safe and without risk to health. Employees must be consulted on any matters concerning their health and safety in the workplace. Health should not be jeopardised by work either immediately or in the long term, including the health of a worker's offspring. At the end of a working day a person should be able to enjoy a full, active social and family life.

The legislation on safety at work in Ireland had its origins in the Factories and Workshops Act 1901, but the first item of modern legislation which controlled conditions in factories and a limited number of other industrial workplaces was the Factories Act 1955, some sections of which are still in force today. This was followed by the Safety in Industry

Act 1980.

The Safety, Health and Welfare at Work Act 1989 is framework legislation relating to all workplaces and imposing general duties on all employers (including the selfemployed), to their employees and non-employees who may be affected by the work activities. Its main features are:

- general principles of safety, health and welfare applicable to all places of work
- general duties imposed on employers, employees, manufacturers and suppliers
- emphasis on the management of safety, health and welfare at work by means of the requirement on all organisations to compile a Safety Statement
- the establishment of an independent enforcement body, the Health and Safety Authority (HSA), with new powers for enforcement
- authorising the HSA to establish technical standards or approved codes to clarify legal requirements
- provision for the replacement of existing legislation with regulations that would apply to all places of work.

Control and inspection

The HSA is a state-sponsored body under the aegis of the Department of Enterprise and Employment. It is the national body responsible for the administration and enforcement of the occupational safety and health system with the objective to minimise the social effects and the economic costs which accrue from accidents and ill-health at work. The 1989 Act also makes provision for the setting up of advisory committees to assist the Authority in its work; six of these were established by 1993. The role of the HSA is most aptly summarised by the functions assigned to it under the Act:

- to make adequate enforcement arrangements
- to promote the prevention of accidents and injury to health at work
- to promote the safety, health and welfare of persons at work
- to provide information and advice
- to ensure that hazards and risks to health and safety are researched.

Enforcement of the law is undertaken by the HSA's inspectors at eight locations throughout the country. It has recently stepped up the number of prosecutions taken in respect of workplace deaths and serious injuries; for example, 39 prosecutions were initiated in 1993. These were concentrated in the employment sectors at greatest risk: manufacturing industry and construction.

The priority for the HSA in 1992 was the development of proposals for regulations to effect over 20 EU Directives on occupational health and safety and dangerous substances. This work involved a fundamental review of many existing statutory provisions and entailed a considerable resource input. Since then the Authority has continued to draft legislation for Government which implements EU Directives and updates national legislation. By 1993 the main priorities of the HSA's programme were:

- to improve the enforcement base
- to implement the Information Technology (IT) enforcement function
- to prepare regulations and guidelines based on EU Directives
- to increase awareness and ensure compliance with new regulations
- to put in place adequate staffing structures.

The advisory committees also continue to bring together organisations with tripartite representation that can contribute to identifying health and safety issues in individual employment sectors.

Two new advisory committees on construction safety and dangerous substances were established in 1992, while a further advisory committee on occupational health and safety in the defence forces was established towards the end of 1993. New committees on the education sector and on mines, quarries and offshore activities were launched in 1994.

The principal aim of the HSA's programme of work for 1994–96 is satisfactory compliance nationally with the Safety, Health and Welfare at Work Act 1989. This is to be achieved through promotion, enforcement, and the provision of advice and information, thereby assuring workers a safe and healthy working environment as well as protecting the public from the adverse effects of work activities.

The detailed elements of the programme of work are organised under the Directorates:

- Information and Training Directorate
- Legislation and Resources Directorate
- Specialised Services Directorate
- Operations Directorate

Other bodies who participate in health and safety include the Government Departments of Health (Environment Protection Officers), the Environment and Social Welfare.

Information and assistance

The overall policy highlights the importance of access to information and advice on health and safety, but it is questionable whether many employers or the trade unions representing employees at workplace level are sufficiently aware of the range and complexities of the issues linked to health and safety. In promoting the implementation of the occupational health and safety system, another core factor is the need to generate the data which would enable employers, government and trade unions to become aware of the economic costs of accidents and ill-health at work. Such awareness is important in recognising the competitive advantages of occupational health systems as well as creating a better work environment.

Training

Training and education is generally regarded as the key to preventing accidents and occupation illness at work. However, even though there is a strong emphasis on education and training, there is no national programme, as yet, on health and safety awareness operating at primary or secondary levels. This weakness in policy is one of the main concerns for the future and will be addressed by the new advisory committee on the education sector which was launched in 1994.

Research

The policy of the HSA is that while it will itself have the capacity to conduct some research, its primary role should be to act as a catalyst, a co-ordinating body and supporter in developing research priorities, providing raw data and access to resources. It should stimulate individuals and organisations to conduct projects which fill important information gaps or otherwise accord with its priorities. Another important role for the HSA is in converting research findings into practical, comprehensible material relevant to employers, workers and their organisations.

3 Health and safety outputs

Occupational accidents

The overall reported figure for accidents (both fatal and non-fatal) in 1993 was 3,606. According to the HSA, there were 64 reported workplace fatalities in 1993, the largest number of which were in agriculture (21). The number of accidents (both fatal and non-fatal) reported to the Authority for 1992 was 2,874, with reported fatalities numbering 46. There was therefore a 20% increase in the number of reported accidents in 1993, as well as an increase in the number of reported fatalities. This may be a result of the Safety Health and Welfare (General Application) Regulations 1993 which require that all occupational accidents involving disablement from normal work for more than three days be reported to the HSA. Injuries to the public as a result of work activities are also reportable if medical treatment is required or where a fatality occurs. A gradual increase in reported accidents is to be expected as many workplaces begin to report for the first time. The analysis of this accident data should enable effective planning aimed at the prevention of accidents and ill-health at work.

The HSA has pointed out that there is a major problem with the under-reporting of workplace accidents and occupational diseases. In order to improve the data base on accidents, a question concerning accidents and ill-health was included in the Labour Force Survey which was conducted by the Central Statistics Office in 1993. Preliminary analysis indicates that around 80% of all workplace accidents and 95% of cases of ill-health caused by work are still not being notified. Information concerning the number of accidents and incidents of ill-health in the principal economic sectors will be available from this survey by late 1994. Sectors of particular concern include the farming, forestry and fishing sector, manufacturing and construction. Reliable statistics on farms are only available in respect of fatal accidents. A survey on safety and health in 1992, however, reported approximately 5,000 accidents giving rise to personal injury occuring on Irish farms each year. Of these over one third resulted in a hospital stay of an average of ten days.

The accident figures quoted above paint a limited picture of the size of the problem in relation to accidents and ill-health at work. The notification of accidents to the HSA is clearly important and the figure of 80% under-reporting must be improved upon. The General Applications Regulations 1993 have been in effect for only a short time and it is clear that many employers are not aware of these legal requirements and are not complying with them.

Occupational diseases

The Occupational Medical Service of the HSA acts as a specialist support unit, in these areas in particular:

- the investigation of occupational diseases
- advice on health surveillance
- advice on preventive systems for particular health hazards.

A total of 28 workplaces were visited by the Medical Advisers in 1993. Areas of concern included sheep dipping, occupational asthma, colophony, three pack paints, the mushroom industry, violence and stress at work, and smoking. Violence has been identified as one of the primary causes of injury to health-care workers. A survey has been conducted on 21 inpatient institutions to assess the extent to which psychiatric hospitals, which are areas of potential high risk, had specifically covered the area of violence prevention in their Safety Statements. In some 71% of institutions violence was included in current or draft Safety Statements.

Absenteeism

At national level little attention has been devoted to the nature, extent and cost of absenteeism. A comprehensive survey on absenteeism in 1992 with a sample of 350 companies, demonstrated that absenteeism from work and the direct costs of such absenteeism is a serious problem in Irish work settings. The relationship between absenteeism, accidents, disease and psycho-social factors has yet to be systematically researched. As there are no official annual statistics on labour absenteeism, its full impact cannot be established and this is a major issue to be addressed in the future.

The economic cost of occupational accidents and diseases and absenteeism

Current priorities include the need to convince industry/employers of the cost-effectiveness of safety management and to estimate the total cost of accidents and ill-health at work to government and employers. In 1994 the HSA launched a survey of the costs of occupational accidents in employment in co-operation with the Northern Ireland Health and Safety Agency. This survey is likely to illustrate the contrast between those companies which implement good health and safety practices effectively and those which place little or no priority on the subject.

4 Assessment of occupational health and safety policies

The National Health and Safety Policy developed by the HSA stresses the importance of Safety Statements and the initiation of safety consultation procedures, including the selection of safety representatives. Despite the development of enforcement activities, however, there appears to be a lack of follow-through. Even where there is a Safety Statement, its provisions are not always put into effect, so that working procedures are inadequate, training is deficient, the workplace is inadequately maintained, or there is a lack of dialogue between employer and employees which might have prevented the accident or fatality. These deficiencies, together with the high under-reporting of accidents and occupational diseases and the widely held feeling that the number of inspectors and inspections is inadequate, contribute to a deep concern regarding the effectiveness of the current occupational safety and health system.

The following core weakness of the Health and Safety System can be identified:

1 A poor statistical base. Without the existence of adequate national statistics it is very difficult to evaluate effectively the health and safety system. It can also hinder future research.

2 A lack of resources. For example, although the HSA places a strong emphasis on developing its range of publications and videos, it acknowledges that it cannot currently match the demand from the public for information.

3 A high rate of under-reporting of accidents hinders the analysis of the causes of accidents in order to prevent future occurrence. It is clear that many employers are not aware of, and are not complying with, new legal requirements.

4 The HSA (and training bodies) put most emphasis on the physiological aspects of health and safety. Although this is an area of extreme importance, many feel that the psychological and social aspects of health and safety should receive equal attention.

5 Other bodies feel that not enough emphasis is being placed on helping SMEs to manage health and safety with a minimum of bureaucracy.

6 Limited resources for research is a current weakness in national policy. What also hinders the development of relevant research is the absence of dialogue between those directly concerned with day-to-day safety and health issues (workers' representatives, policy-makers, employers, etc.) and those who are carrying out research or are otherwise involved in the selection of critical issues for research. A comprehensive database on the work environment and quality of work is urgently required.

The gaps outlined above are all regarded as important areas for future consideration. Other main concerns include:

1 Lack of government funding – increased funding is required to develop the resources of the HSA (e.g. enlarge the inspectorate), to enhance information and training at national, company and school levels, and to support research on health and safety.

2 Health and safety, especially training, may appear low on trade union agendas due to current economic pressures and the need to create new jobs to reduce unemployment. This clearly means avoiding unnecessary costs, but it should not be to the exclusion of health and safety training. The HSA continues to be convinced that the prevention of many workplace accidents is achievable at little or no additional cost to the company and will in fact enhance competitiveness.

3 The recession which involved pay-roll cutbacks in industry is slowing up the implementation of safety and health programmes.

Priorities for the future include the need to build occupational health and safety into management programmes and make resources available, and the need to study the implications of new technology on occupational health and safety (particularly ergonomic and psychological implications). In the future, psycho-social criteria of stress and wellbeing at work will need to be actively developed and integrated into a comprehensive model of health and safety, and a framework for implementation of policy put in place. Although the recession has had a major negative effect on the development of occupational health and safety and the enforcement of legislation, the recent improvement in the economy is resulting in a demand for occupational health and safety professionals.

The role of the European Union

All the bodies interviewed for this study agreed that the EU has a significant impact on health policies in Ireland, with the harmonisation of the market having had positive effects on occupational health and safety. The tripartite approach is a driving force in facilitating a realisation of EU Directives and regulations. Occupational health and safety regulations based on EU Directives and EU standards and codes of practice are setting occupational health and safety standards for Irish industry. The Social Charter has had a major impact on Irish legislation. The EU is an issue in that it may be perceived as undesirable that policies are 'coming from the EU', rather than adopting a more independent approach. However, on the whole the feeling in Ireland is that the EU is a positive influence, where it matches the societal ethos.

Ireland, given its size, its limited resources and its high unemployment levels, will be faced with a major challenge to balance the drive towards increasing employment with putting health and safety regulations in place, particularly in sectors that are by definition vulnerable (e.g. SMEs, farming, etc.). Given this socio-economic context, Ireland has made significant strides in introducing all of the Directives under the EU Framework Directive (89/391/EEC). Employers point out that the present developments, legislation and structures with regard to health and safety have been steadily evolving over the past 12 years with a strong tripartite involvement. This development has interfaced with EU priorities.

There is a feeling amongst employers that there is a need for greater practical guidance from the EU on implementation. The present onus on employers to develop practical modes of implementation is heavy. Many Irish workplaces are SMEs and an analysis of the practicality of the measures emanating from the EU needs to be undertaken. This latter concern is viewed as an urgent priority. According to the report of the Task Force on Small Business, small-scale enterprises cover 90% of Irish workplaces. In 1993, the HSA conducted a study of 250 SMEs employing less than 50 people, which showed that 77% of SMEs had no Safety Statement and that 85% did not have safety representatives. This study has been extended to 2,500 SMEs in 1994 to monitor awareness of health and safety and to determine what are the best routes for disseminating information to these SMEs. In addition, a safety pack for small companies was recently introduced by the HSA.

From the perspective of HSA representatives, the EU has had a massive impact on recent health and safety legislation and policies in Ireland. The trade unions view the EU as a major actor in the evolution of health and safety issues. The EU Advisory Committee on Health and Safety is seen as providing an important learning ground for transnational perspectives, problems and practice. Ireland responded very favourably to opportunities provided by the European Health and Safety Year and is viewed as being very much in tune with EU developments and initiatives.

1 The context of national occupational health and safety policies in Italy

Economic structure

There are just over 19.8 million people in employment in Italy, representing 33% of the general population. Employment is most prevalent in the services sector (60.2%), while about half as many people are employed in industry (32.7%) and the remainder (7.0%) in agriculture. There are significant differences between the economic and social structure of the northern, central and southern regions. Just over 50% of employed people are found in the north, with 29% in the south and 20% in the central region.

The labour market

Unemployment

There has been a rising trend in unemployment. Between October 1992 and October 1993, for instance, unemployment rose from 5.4% to 6% of the population as a whole. More recent figures show that this trend continued, with a further 385,000 employees losing their jobs between January 1993 and January 1994. The percentage of the workforce unemployed by 31st January 1994 was 11.3%, comprising 8.7 % of the male workforce and 15.7% of the female workforce. Since 1988 there has been a general decline in the number of young people in employment in all the geographical regions. At the end of August 1993 there were 449,765 young people in work compared with 556,506 in 1988.

Migrant workers

Official data on migrant workers suggests that there were 76,988 such workers in Italy in 1993, although it is acknowledged that this figure greatly underestimates their actual numbers. Such underestimation occurs because the employment of the majority of migrants is precarious and often illegal and therefore not recorded.

Industrial relations

There is currently a tendency towards a more collaborative approach in industrial relations between trade unions and employers' associations, in contrast to the 1970s and early 1980s. Such collaboration is not yet operating fully in collective bargaining, however, especially not at branch and company level. In working towards this collaborative approach, the social partners have been influenced by developments in European Union legislation.

Provincial-level joint safety and health committees only exist in a few industrial sectors (such as construction). Joint health and safety committees at company level are provided for by agreements in some industrial sectors (in the metal and chemical industries, for example). In contrast, it is not unusual to find worker health and safety experts as members of workers' councils in enterprises employing more than 15 workers. The trade unions are particularly interested in the introduction of the provision of safety delegates that will be instituted by the law implementing the EU Framework Directive 89/391/EEC in Italy. Generally, the European dimension in legislative change is considered a positive factor by both employers and trade unions.

The European dimension

There have always been delays in the transcription and implementation of European Directives in Italy. The first adoption took place in 1991, with DL 277/91 which implemented five Directives concerning amongst other things workers' protection against noise, asbestos and lead; it also anticipated some of the contents of the Framework Directive 89/391/EEC. DL 277/91 introduced for the first time the practice of risk assessment in the specific areas covered by these Directives. As a result, three years since the implementation of this legislation, all big companies, as well as a large majority of medium-sized and small companies, have now, for example, carried out a technical evaluation of the noise exposure of their employees.

The rapidly changing political situation between 1990 and 1993 caused a delay in the adoption of other relevant Directives, including the Framework Directive, which should have been implemented before 20th September 1994. However, a decision of the High Court has established that European legislation shall be applicable when specific national legislation is lacking, or is less protective.

In 1993 a referendum established the separation of environmental controls from the public health services, giving the opportunity of instituting a national agency and local services for environmental protection. In the same year another new law greatly modified the organisation of the National Health Service and established a Prevention Department in every health district to take responsibility for the occupational health services (OHS), public health services and veterinary services. It is intended that the new arrangement will be fully developed by early 1995.

2 Occupational health and safety policies and structures

Legislation

Italian health and safety legislation is largely based upon decrees enacted during the 1950s that are still used today, although they have been subsequently augmented by a large number of specific additional provisions, and more recently (November 1994) by the adoption of DL 626/94 which implements some European Directives including the Framework Directive 89/391/EEC. The legislation on health and safety at work relates directly to the constitution which considers health a basic right that should be achieved by taking the preventive measures that are technically feasible regardless of the economic costs involved. The recurrent general principle in the legislation (supported by criminal sanctions) is that employers are responsible for taking these technically feasible measures to reduce risks.

The main law concerning occupational disease and accident insurance was passed in 1965. The list of recognised occupational diseases was last modified in 1994 when the number rose to 50 in industry and 27 in agriculture. Diseases that are on this list are recognised without there being any need to prove their occupational origin. For other diseases to be accepted as being of occupational origin, the link has to be proved.

Occupational health and safety structures

In 1978, the Health Services Act (Law 833/78) instituted in every district local health units (USL) whose staff include doctors, nurses, safety experts and industrial hygienists. The law devolved some of the powers of the Labour Inspectorate to the USLs, thus providing them with the regulatory powers in health and safety traditionally associated with labour inspectors and enabling them to use these powers in the workplace in addition to their preventive advisory activities.

Since 1981, however, the mission of the OHS has not simply been to ensure compliance with the law, but also, and most importantly, to promote positive changes in the workplace. Achieving compliance with the law is only one of the ways of creating a better work environment and, as a consequence, improving the condition of workers' health. For this reason, inspection is only one of the tools used by OHS personnel, who normally define themselves with the titles of their specific professions, such as: occupational doctors, industrial hygienists, chemists, engineers or nurses.

Control and inspection

The approach of the OHSs to intervention is by means of prevention plans which are drawn up on the basis of an evaluation of risks. The aim of such intervention is to assess the level of risk, to control factors affecting the health of workers and, most importantly, to find ways to improve conditions overall. In the intervention schedule, inspection is always the first step; this is because compliance with the law is regarded as the starting point which offers workers the minimum level of protection.

If serious deficiencies are found during inspection, OHS personnel can require the employer to adopt immediately the protection measures prescribed by law, as well as measures that are not prescribed. In the latter case the employers have recourse of appeal to the President of Regional Government. In the event of serious danger, OHS personnel can stop the work and keep machinery, or even the workplace, in sequestration until a judge's decision.

There are no national statistics on inspection activities in Italy. In 1988, however, SNOP (the National Association of Prevention Professionals) conducted a national survey which concluded there were about three thousand professionals working in OHSs, more than two thirds of whom were based in the northern and central regions.

Data published by Emilia Romagna Region referring to 1991 show that in this region there were some 551 professionals, in the main comprising doctors, biologists, chemists, engineers, nurses, administrators and technicians. Of these the largest group were technicians (37.7%), followed by doctors (20%) and nurses (18.1%). In 1991, in collaboration with these personnel, the OHSs in Emilia Romagna made prevention plans in 9,271 companies employing 347,852 workers, 191,725 (55%) of whom were assessed as exposed to some risks. The data on Emilia Romagna region's activities can be taken as representative of the northern and central regions, but the situation is totally different in southern Italy, with some exceptions in Puglia.

Another important activity carried out by the OHS is the evaluation, prior to building, of the construction plans of factories. Municipalities require sight of such plans before authorisation and building approval is granted.

OHS personnel also make medical visits as part of Medical Survey projects, or when asked jointly by workers and employers to perform the medical assessments provided by industrial hygiene laws. In 1991 medical assessments were performed in 2,589 companies, involving 43,984 workers in Emilia Romagna.

With regard to the penal sanctions against employers as a result of accidents or occupational diseases, in 1991 there were a total of 2,284 actions brought against employers concerning serious and fatal accidents and occupational diseases, the majority of which were the result of accidents (1,886). Just over half of all the actions were brought by the OHS (1,168) and just under half by judges (1,066) with the remaining

50 actions being brought by others.

Company occupational health services

At present there is no law that establishes the duty to organise private (company) occupational health services. Legislation (confirmed by Law 626/94 which implements the Framework Directive 89/391/EEC) obliges employers to make workers available for visits made by a so-called 'competent doctor' who, as established by DL 626/94 (and before by DL 277/91), must be a specialist in occupational health. DL 626/94 does, however, establish the duty to organise safety services. Consequently, only large companies have their own occupational health services and safety services; these structures therefore cover only a small percentage (about 10%) of employees. In the case of hearing protection, however, the provisions of DL 277/91 increased the number of workers receiving medical surveillance. As a result, in northern Italy, currently 40% to 50% of workers exposed to risks are covered by medical assessment.

Information and assistance

In terms of publications, there are several reviews directed at different targets. The main (and oldest) review is *La Medicina del Lavoro* (Occupational Health), published every three months by The Occupational Health Clinic in Milan, with a circulation of 1,000 copies. Until 1993 the trade unions published a review called *La Medicina dei Lavoratori* (Workers' Medicine), which was a scientific review publishing experiences in which workers (including workers' councils, trade unions, etc.) were actively involved. It also had a circulation of about 1,000 copies. The SNOP has published its own review, entitled *SNOP*, since 1985, with a circulation of 6,000 copies. Several regional documentation centres jointly publish monthly, *Lavoro e Salute* (Work and Health), which disseminates brief information about local experiences, national and international projects, publications, congresses and seminars. The circulation is 11,000 copies.

Since 1982 the region of Emilia Romagna has published a series of booklets on life and workplace prevention called *Contributi*, better known in Italy as 'green booklets' because of the colour of the cover page. About 3,000 copies per issue are sent to all the relevant institutional bodies. More than 40 green booklets have been published, making an important contribution to the work of the occupational health services in the whole country. The region of Tuscany publishes a series called *Sicurezza Sociale* (Social Security) with similar contents to those of the green booklets.

There are also other publications, printed by private institutions and supported mainly by producers of safety devices.

Training

Whilst there are a a great number of courses and seminars for private and public professionals, especially doctors, organised by private and public institutions and scientific associations, workers and employers have little possibility of obtaining real and effective training on prevention and safety at work. Recently, however, employers' associations and trade unions have started jointly to organise several courses and seminars, often with the collaboration of the public occupational health services.

Economic incentives

No specific economic incentives are provided for companies that adopt safer or healthy measures in workplaces. The range of penalties varies; it was raised significantly following the implementation of DL 277/91 which introduced a new and higher range of fines. Employers can be sanctioned with fines of up to 25,000 ECU for serious omissions, such as not having a risk assessment, failure to adopt 'concretely feasible' measures to reduce risks, failure in providing individual protection devices, and failure to adopt measures indicated by the inspection and control authorities.

Research

Research in occupational health is performed mainly by universities and the Higher Institute for Work Safety and Prevention (ISPESL), chiefly in relation to medical problems.

Occupational health insurance

Occupational health insurance in Italy is organised by a public institution, the National Institute for Insurance of Occupational Accidents and Diseases (INAIL), which is responsible to the Ministry of Labour. Its General Director is nominated jointly by government, trade unions and employers. INAIL also makes provision for disabled workers. Insurance is compulsory for all enterprises in the industrial, agricultural and services sectors. It is financed through employers' premiums, which differ between sectors depending on levels of estimated risks, and on the accident and disease rates of individual factories or workplaces.

3 Health and safety outputs

Occupational accidents

In 1991 there were 1,020,000 accidents reported (790,000 of these were in the industrial sector, and 230,000 in agriculture). There were 1,858 fatal accidents (1,058 in the industrial sector and 800 in agriculture). The number of accidents have tended to rise in recent years; however, INAIL estimates that about 15% of accidents are unreported.

Occupational diseases

The recorded incidence of occupational diseases between 1985 and 1989 shows fluctuations of between 20,000 and 30,000 recognised notifications per year. An increase in notifications in 1988 was the result of the action of the High Court which made it possible for diseases not on the list of recognised occupational diseases to be included, if their occupational origin could be proved. Nearly 50% of the recognised cases were in relation to noise, with diseases caused by skin irritants (nearly 20%) being the next largest category, followed by diseases caused by silica (just over 6%) and other dusts (just under 6%).

Although notifications of occupational diseases grew during the 1980s, this can be explained by the improved awareness of the OHSs which as a consequence brought to light a number of 'hidden' diseases, resulting in increased notifications. In more recent years a decrease in 'traditional' diseases, especially in industry, has been registered. At the same time as awareness of new work-related diseases in traditional sectors (industry) grows, there continues to be a problem in the diagnosis and recognition (and therefore compensation) of 'old' diseases in agriculture, which are still not included in the official list of recognised occupational diseases.

The economic cost of occupational accidents and diseases and absenteeism

A number of attempts have been made in recent years to assess more accurately the economic cost of work accidents, occupational diseases and absenteeism. A broad study based on 1991 data estimated the total cost of accidents to be in the order of 22.84 billion ECU, which represents about 3.05% of GNP. The cost of the public prevention services in Italy, however, is calculated to be about 1.5 billion ECU, or 3% of the total NHS cost. The trade unions and scientific associations have proposed that this amount should be increased to 6% (3 billion ECU) in order to increase the effectiveness of prevention and, indirectly, to lower the economic cost of accidents.

4 Assessment of occupational health and safety policies

The period covered by this review is particularly interesting with regard to health and safety policies. The debate about the need to make institutional changes in the organisation of public intervention in the field of health and safety, which started some years ago, was concluded positively at the beginning of 1994, with the approval of the law that will reorganise the National Health Service and establish a Prevention Department with occupational health services in each local area.

At the time this report was being put together, the trade unions and the majority of health and safety experts were concerned about the possibility of delays in the reorganisation and implementation of European legislation caused by the new Government led by right-wing parties, and about the possibility that this Government might change the general approach to health and safety at work. Later changes in Government suggest that such fears may have been unwarranted. Employers, as shown by the declaration of the President of Confindustria (the employers' association), believe that the implementation of EU legislation is a priority, and the Minister for the co-ordination of EU policies has said that the implementation of EU legislation must take into account any higher levels of protection in existing legislation.

There are several critical points that need urgent intervention:

1 Unemployment, migrant workers (often illegal) and precarious employment can affect the improvement of working conditions and even worsen them.

2 There is an urgent need for up-to-date and more co-ordinated legislation; in this respect the speedy implementation of EU Directives is very important.

3 Public occupational health services must be organised at national level which requires resources and personnel. In the meantime employers must be more actively encouraged to set up their own occupational health and safety services at company, branch or local level. Public and private services should be encouraged to collaborate in providing information as well as the assistance and training of workers, employers and experts. This should be a priority task for public services.

4 There is a need for further studies and information about accident and disease trends.

5 Specific and effective campaigns to reduce work accidents and occupational diseases, especially in critical sectors (like construction and agriculture) should be conducted, using, on the one hand, the powers of control and inspection, and, on the other, information and education, including the mass media.

6 Universities and technical schools should be made aware of the need to assess new situations, technologies and practices in order to ensure the availability of effectively trained professionals and experts.

7 A specific policy on prevention in occupational health and safety should be adopted by Government, resulting in the implementation of programmes to improve working conditions.

1 The context of national occupational health and safety policies in The Netherlands

Economic structure

The Netherlands has about 15 million inhabitants. The labour force in 1992 was about 6.5 million, of which 60% was male and 40% female. As many as 98% of all companies were small to medium-sized (1–100 employees), with large companies concentrated in manufacturing and other services.

The service sectors employed the largest number of people, accounting for 66% of the total workforce, with only about 25% of the workforce employed in manufacturing and construction. About 487,000 employees (8% of the total workforce) work in the public sector, which has 26% of the female workers; the private sector has 41% of the female workforce.

The labour market

Unemployment

The number of people receiving unemployment benefit fell between 1987 and 1992 due to a fall in unemployment amongst males and younger age-groups, altough there was an increase amongst older workers. In 1992 the official unemployment rate was about 5%.

Temporary employment

In 1991 a total of 141,000 jobs (about 3%) were held by temporary workers. Disproportionately more women are employed on a temporary basis, although temporary jobs are full-time or substantial part-time (20 hours or more). Temporary work is mostly found in industry, and the trade and service sectors.

Part-time work

Part-time work has become an important employment opportunity for three demographic groups in Dutch society: younger, older and female workers. Services and trades rely most heavily on part-time workers. In 1987 about 15% of male employees worked 35 hours or less; for females, however, it was 60%. In the years 1987–1992 the number of part-time jobs increased by 29%, whilst the number of full-time jobs in the same period increased by only 4%. A notable increase in part-time jobs occurred in commercial services where there was an increase of 37%, whilst the number of full-time jobs in this sector increased by 16%.

Moonlighting

According to a Government report, about twice as much unpaid as paid work was carried out in the 1980s. Paid work was mainly undertaken by men (75%) and unpaid work mainly by women (also 75%). The information on informal work– i.e. illegal work, or legal work without paying taxes – is limited and not very reliable. The Central Bureau of Statistics estimates that informal activities are responsible for 10–15% of the national income, with a substantial increase of informal work likely in the future.

Male and female workers, young and older workers

The Dutch working population is steadily ageing and the percentage of women is increasing compared with men. According to studies, there are distinct differences in the quality of work between older and younger workers, and between men and women, so that younger people, and men in particular, carry out work that involves risks to their health, safety and well-being, whilst women tend not to be able to use their education in their jobs and have fewer opportunities for promotion.

Migrants

A total of 448,000 migrants (non-nationals as well as people born outside the Netherlands) were employed in 1992, which is 7% of the total working population. The sectors with the highest proportion of migrant workers are industry (25%), trade, hotels and catering (18%), commercial services (15%) and other services (31%).

Industrial relations

Levels of collective bargaining

The industrial relations system in the Netherlands is highly centralised and there is much emphasis on creating a consensus between the social partners and national government. The position of trade unions in individual companies is rather weak; however, the central organisations of employers and employees, which operate on a national level, carry a great deal of weight in Dutch politics. Labour relations are harmonious and serious conflicts are rare. However, there is a tendency for decentralisation towards sector and firm level.

Workers' councils

Under the Workers' Council Act it it is compulsory for companies with 35 or more employees to set up a workers' council. Companies with fewer than 35 employees are not compelled to do so, but the law regulates a limited participation on the part of all employees. although the participation of employees in a firm with fewer than 10 employees is not regulated by law. The Act also covers some of the rights of employees in the field of working conditions. An example is that the employer has to write an annual policy plan for the safety, health and well-being of employees, which has to be discussed with the workers' council. Under the Working Conditions Act, the employer also has to discuss the choice of occupational health service with the Workers' Council. The Workers' Council has the right to accompany labour inspectors on company visits.

Membership of trade union and employer organisations

One quarter of all employees (1.8m) belong to a trade union. Employees are organised in sectors which fall under central national union organisations. The most important ones are: the social-democratically orientated FNV (Federation of Dutch Trade Union), the CNV (National Christian Trade Union) and the smaller MHP (Trade Union Federation for Middle and Higher Personnel).

The most influential (and larger) employers' associations are the VNO (Federation of Dutch Industries) with 25,000 members and the Christian NCW (Dutch Federation of

Christian Employers) with about 45,000 members. Recently these organisations and the organisation representing employers in small and medium-sized enterprises (SMEs) merged.

2 Occupational health and safety policies and structures

Legislation

The Working Conditions Act

The most important piece of legislation with respect to the quality of working life is the Working Conditions Act which was passed in 1980, although it took ten years to be fully implemented. It aims to increase levels of safety in the workplace and to improve the physical and mental health and well-being of employees. Since 1990 all persons working for an employer are covered by this Act, both in the private and public sector, and in large corporations as well as in small firms. The Act is primarily an enabling act: it is essentially a framework which forms a basis for more detailed decrees. Another important aspect of the Act is that it defines the role of the Minister, the Labour Inspectorate, employers, employees, Workers' Councils and safety and health specialists.

Self-regulation is an important concept underlying the Working Conditions Act. Responsibility for safety, health and well-being at work is generally placed with the employer: 'In organising work, the method of production and work adopted must be such that the highest possible level of safety and health protection is attained and people's well-being is improved'. To establish these conditions the employer has to co-operate with the employee who also has obligations, such as to follow instructions and training with respect to safety procedures. A committee is currently preparing proposals which may lead to the deregulation of the Act.

The Working Conditions Act was amended on the 1st January, 1994 as a result of:

- the coming into force of the European Framework Directive 89/391/EEC
- new Dutch legislation connected with preventing sickness absenteeism and

work disability
- the desire to create a legal basis for certifying the occupational health services.

The major changes are that all employers are obliged to:

- assess the risk within the working environment
- make clear what measures are to be implemented to decrease risks in the working environment
- implement a policy on sickness absenteeism (prevention, control and guidance)
- engage professional help on matters of health, safety and well-being; larger firms have to have an official contract with an occupational health and safety service.

Related legislation

As well as the Working Conditions Act, there are several further relevant laws relating to working conditions including:

- several Acts in safety and health (mining, home work, dangerous work, pesticides, etc.)
- Acts on working hours
- The Workers' Council Act
- The Employee Insurance Act (sickness absenteeism and work disability)
- Acts on preventing sickness absenteeism and work disability
- Environmental legislation.

Control and inspection

Mission

The Labour Inspectorate inspects companies to ensure that the Working Conditions Act is being observed. If companies fail to comply with the legal standards, sanctions can be imposed.

At the end of the 1980s, the Labour Inspectorate was repeatedly criticised for its enforcement practices. The main criticisms were that inspectors did not visit enough companies, that they went into too much detail during inspections and that the inspectors did not concentrate on systematically inspecting the general work environment policy of the company. Partly in reaction to criticism about its inspection practices, the Labour Inspectorate amended its policies in the early 1990s and published a broad-based policy programme selecting different approaches for different branches of industry, with priority given to the high-risk branches. Inspectors now focus on checking the procedures set up by employers to ensure good working conditions, as well as checking that the working conditions requirements are fulfilled.

The Labour Inspectorate can carry out enforcement procedures and it can also act in cases of conflict between employer and employees. In general, however, its approach is to stimulate consultation and co-operation between employers and employees and to stimulate organisations to develop policy on working conditions, safety and health.

Structure

The Labour Inspectorate is organised in 10 districts covering the whole of the Netherlands. Some 300 inspectors are supervised by the central service in The Hague known as the Labour Directorate-General which is part of the Ministry of Social Affairs and Employment. It serves as the policy department of the Labour Inspectorate.

Field of competence

The Labour Inspectorate has contacts with both employers and employees' representatives in workers' councils. Employees have the right to contact the Labour Inspectorate without their employer's consent and they are entitled to accompany the Inspectorate's representative on his inspection tour. In case of conflicts, employer and employee have to find their own solution, and only if this proves impossible can both parties apply to the Labour Inspectorate. In this case, the Inspectorate has to hear both parties and look into the problem itself before coming to a decision. The Inspectorate has the authority to instruct employers to follow the provisions of the law and it can make reports for prosecution. In particularly serious cases, if the life or health of employees is in danger, the Labour Inspectorate is empowered to order an immediate stoppage of work. In the event of immediate danger, an employee has the right to stop his/her activities on condition that the situation is reported immediately to the Labour Inspectorate. In such a case, the Inspectorate acts as arbitrator.

Annual coverage of inspections by sector

The number of visits to companies in 1990 was 64,401 (11% of all companies). Every year the Labour Inspectorate receives over 1,000 complaints about working conditions. It investigates about 2,000 accidents annually and an increasing number of reports of occupational diseases (702 in 1990).

Rate of inspectors per worker

About 300 inspectors cover 6.5 million employees in nearly 600,000 companies, giving a rate of one inspector for every 16,667 workers.

Number of sanctions per company

In 1990 the Labour Inspectorate found 35,000 faults in the 64,401 firms visited. With respect to 5,000 of them, the inspectors required substantial changes, and in 400 cases an offence was reported.

Occupational health services

Types of services

Three types of occupational health services (OHS) are distinguished in the Netherlands.

'Joint occupational health services' are non-profit-making bodies working for several firms and administered by management and labour representatives from these firms. In 1989 there were 49 of these services, delivering care to some 4,000 firms with nearly 1 million employees. Since the Government decided that all employers have to engage professional help on safety and health, the number of this type of service, as well as their profits, has increased significantly. The current extent of these services is not known.

'Single occupational health services' work for one firm only. There are some 70 of these serving more that half a million employees.

A third type of service is formed by occupational health care departments of government and local authorities delivering care to their own staff. Nearly all these services are joint in the sense that they serve more than one workplace. There are 66 of them, serving some 3,300 organisations/agencies and more than half a million employees.

Financing

OHSs are paid by the companies which buy the service on an activity basis.

Type of personnel

All occupational health services are required to employ at least the following four experts: an occupational doctor or a social insurance doctor, a senior safety expert, an occupational hygienist and a (psycho-social-oriented) work organisation expert. In 1992 there were 1,091 occupational doctors, of whom 691 were registered occupational doctors and 400 were assistant doctors. In addition, 150 occupational hygienists, 806 paramedics, 513 occupational health nurses and 667 other employees were working in OHS. Thus the total number of employees in that sector was 3,227.

There are about 100 full-time and 150 part-time ergonomists employed in Dutch industry, most of them not working in OHS. There are, also, some 1,400 safety experts employed in firms and institutions. Furthermore, there are some 1,000 social insurance doctors employed, who are consulted in relation to the Sickness Benefit and Work Disability Acts. The number of these insurance doctors has decreased dramatically as a result of changes in the extent of the coverage of the Sickness Benefit Act in 1994.

Coverage of the workforce

In total, more than 2 million employees (42% of the workforce) are served by the OHS. This coverage is expected to rise considerably because of the obligation in EC legislation to broaden OHS to all workers. Until 1994 the coverage of occupational health care was the highest in the construction industry, with 100% of employees having access to it; for industry as a whole it was 40% and in services it was approximately 30%. All care is paid for exclusively by employers.

Tasks and activities

The Working Conditions Act requires that, as a result of the Framework Directive (89/391/EEC), every employer is assisted by a certified OHS unit; alternatively, the employer can organise his own services unit to carry out four main tasks:

- health and safety risk assessment
- periodic occupational health examinations for employees
- occupational health consultations for employees
- social-medical guidance of sick and absent employees.

Although the Working Conditions Act establishes these four main tasks, many OHSs spend most of their time on pre-employment medical examinations when workers are recruited, as well as the periodic medical examination of workers exposed to certain risks.

In the private sector, absence certification is carried out by the Industrial Insurance Associations. In the past, certification of sickness absenteeism had also been a task for the OHS, but that is no longer the case.

Link between occupational health services and general health system

There is no direct link between the OHS and the general health system, and the separation between treatment for sickness absenteeism and other treatment has caused occupational doctors to complain that general practitioners and medical specialists are not interested in work-related illnesses. Many employers have also complained that the length of time sick employees have to wait for specialist or hospital treatment adds to the costs they have to bear from sickness absence.

Information and assistance

Legislation is published in the State Gazette. The Ministry of Social Affairs and Employment publishes information sheets, studies, guidelines, periodicals, etc. and has a central department in charge of public information. As well as government publications, much information is published by private publishing houses and through a number of independent periodicals.

Training

Occupational doctors are trained on post-graduate courses for three or four years; there is provision for 25 training days a year. Training for safety and health specialists and labour inspectors is provided, as is training for work organisation experts and occupational hygienists.

Economic incentives

Within the framework of the Sickness Benefit Act and the Work Disability Insurance Act, the Ministry of Social Affairs and Employment recently (1994) issued new legislation directed at lowering the rate of sickness absenteeism and work disability. The measures are heavily based on economic incentives and were mostly introduced in 1993 and 1994.

The measures directed towards employers are:

- the six weeks of sickness absenteeism will no longer be covered by the Sickness Benefit Act but by the employer; this period is two weeks for small companies
- there will be more differentiation between firms in the premiums for sickness benefit paid by firms based on their actual sickness absenteeism rate
- a financial reward for firms that take on a partly disabled person as an employee and a financial penalty for firms that allow an employee to become disabled.

Other measures, directed towards the employee, are:

- 70% instead of 100% payment in the first six weeks of illness
- people under the age of 50 who are declared disabled for work will receive much lower benefits than was the case previously
- the concept of 'suitable work' is broadened so that disabled people will have to accept jobs that are offered, even if the level is below that of the job held; the sanction against not accepting such a job is loss of benefits.

Research

The general situation

In the Netherlands there is no central research fund for the whole field of quality of working life. Nevertheless, central government (the Ministry of Social Affairs, Labour Directorate-General) employs an active policy of directing and granting research in this area. There are several important research institutes and groups engaged in research in the field of work and health.

Although there are no formal networks, from time to time the Government forms ad hoc working groups and advisory bodies to represent the social partners and experts which then advise the Government on policies.

A certain amount of research is done for the EU (Directorate General for Social Affairs and Employment of the European Commission) and for the European Foundation for the Improvement of Living and Working Conditions in Dublin. Universities tend to draw on their own funds and they sometimes also carry out studies for the Ministry of Social Affairs or branches of industry.

Field covered

The main funding agency with respect to research in the field of work and health is the Ministry of Social Affairs and Employment. Its Research Programme 1994 includes research projects on labour market, social security and working environment issues.

Occupational health insurance

Mission

The Sickness Benefit Act and Work Disability Insurance Act apply, irrespective of the causes (work- and non-work-related), to temporary and permanent incapacity respectively.

Sickness absenteeism

The definition of sickness absenteeism includes any (accepted) claim under the Sickness Benefit Act. In practice, this means that almost any reported case of incapacity for work due to ill-health can be considered as sickness absenteeism. The scope of this definition is very wide and includes illness, minor conditions, industrial accidents and maternity leave.

Work disability

Whereas the maximum duration for sickness absenteeism is one year, the definition of work disability includes permanent incapacity for work (after one year of sickness absenteeism), again irrespective of its causes. The coverage of disability insurance benefits, however, is more stringently defined than that of sickness benefits. The degree of disability is determined by measuring an applicant's 'earning capacity', i.e. the amount a disabled person would still be able to earn in commensurate work, expressed as a percentage of the income earned by healthy, but otherwise similar, persons.

The administration of the work-related social insurances (sickness benefits, disability insurance) is delegated to 15 insurance boards representing different branches of industry. These 'Industrial (Insurance) Associations' are managed by representatives of employer organisations and trade unions; they are supervised by the Social Insurance Council. The Minister of Social Affairs and Employment determines how the burden of social insurance contributions will be distributed between employers and employees.

Recent developments
The topic of absenteeism has been a controversial issue during recent years. A number of changes have been introduced to try to lower the rates of absenteeism. The most recent Government proposals, if introduced, will extend the period during which the employer has the responsibility for the payment of sickness benefit from six weeks to a full year. If this is carried through, it will mean the end of the Sickness Benefit Act.

Monitoring
Apart from the periodic Work and Life Situation Surveys of the Central Bureau of Statistics, there is no specific monitoring system on a national scale to identify the risk factors and risk groups in the working population. Nevertheless, monitoring is now an important aspect of the general policy programme of the Ministry of Social Affairs and Employment, and a key feature of the work and health programme.

3 Health and safety outputs

Occupational accidents
The number of accidents per 1,000 employees fell from 60 in 1968 to around 27 for men in 1989. For women the rate per 1,000 fell from 9 to 5. Although the registration system is not very reliable, it can be concluded that the number of accidents has declined. With both sexes a fall in the number of occupational accidents can be observed with increasing age. The percentage of victims younger than 25 is much greater than can be expected from the age distribution in the working population.

There are significant differences in the risk of occupational accidents between industrial sectors. The risk of accidents is greatest in the construction industry (52 per 1,000 employees), followed by agriculture and fisheries and the manufacturing industry (34 and 28 respectively per 1,000 employees). The risk is lowest in 'other services' (excluding banks and insurances), where it is 3 accidents per 1,000 employees.

Occupational disease

Article 9 of the Working Conditions Act came into force on the 1 January 1988. It broadened the definition of occupational disease to encompass disorders resulting from physical demands (such as musculoskeletal disorders) and mental demands (such as stress).

Since there are no financial incentives for employers or doctors to report occupational diseases, it is impossible to ascertain accurately their occurrence on the basis of the official data. It has been estimated that they may be under-reported by 20%. According to official data, occupational skin disease forms 98% of all reported occupational diseases from both men and women. A large number of such cases are linked to accidental events, such as burns and the acute consequences of the inhalation of chemicals causing lung oedema.

Absenteeism

Sickness absenteeism

The rate of sickness absenteeism has increased from 3% to 7% in the period 1953–1992 (with a peak of 8% to 9% in 1978–1979). One explanation for the increase is the changed composition of the labour force. Since the beginning of the 1960s, more married, slightly older women with children joined the labour force. Another explanation seems to be that people tend to report sick sooner than before (as a result of better health-care possibilities and changing values regarding work and leisure). A further explanation is that there has been an increase in the total work-load which has not been accompanied by a sufficient increase in the time for recuperation.

As a consequence of the changes in the Sickness Benefit Act at the beginning of 1994, sickness absenteeism decreased artificially in the official data (to about 4%). Data show that sickness absenteeism rates increase with age: the rates for older employees are about 3% higher than those for younger employees. In addition, female employees in all age categories show higher absence rates than male employees.

Industrial sectors with a high rate of absenteeism (about 9% in 1992) are: the building industry, the health services, the cattle and meat sector, the timber and wood industry, hotels and catering. A low rate of absenteeism is to be found in merchant services, banking and finance, bakeries, harbours and dairy work (5–6%).

Work disability

According to recent data, the number of people declared disabled for work totalled 882,000 by the end of 1990 out of a workforce of 6.5 million. The number of newly disabled people in the private sector increased from 15 per 1,000 insured in 1970 to about 23 per 1,000 in 1977–1981 and then dropped to 13 per 1,000 in 1992. Two diagnoses stand out above the others: musculoskeletal disorders and psycho-social and mental disorders.

Women nowadays have the same chance as men of being declared disabled. The chance of being declared disabled increases with age. Statistics show that at least 3% of people who are more than 45 years old run a chance of being declared disabled.

There are significant variations in the number of people newly disabled for work according to industrial sector. Over all industrial sectors (private and public) the figure is 17 per 1,000 insured in 1992. Sectors with a high number of newly disabled people are: cattle and meat (24 employees per 1,000 per year), building, hotels and catering, and health services. A low number of newly disabled people can be found in banking and finance, harbours, and agriculture (12 to 15 per 1,000 in 1992).

The economic cost of occupational accidents and diseases and absenteeism

In 1990, for a working population of about six million, approximately 4.82 billion ECU were spent on sickness benefits and approximately 10 billion ECU on disability payments. Not included in these figures are the costs of the administration of benefits and the other costs of sickness absenteeism such as productivity losses and/or the costs of using temporary replacements.

4 Assessment of occupational health and safety policies

Main concerns in relation to health and safety

Two contrasting strands of development can be discerned in the data on the quality of working life in recent years. Developments were relatively favourable with respect to some working conditions, such as vibration and shock, noise, dangerous and dirty work; but they were relatively unfavourable with regard to the mental and physical side of work - work pace, relationship between education and job content, and work-load.

The health of the working population appears to have improved from the mid-1970s to the 1990s. There is a favourable development in several health indicators (the evaluation of one's own health in health surveys, the use of medication, sickness absenteeism rates, occupational accidents, work disability rates) during that period. It is possible that the decrease in some risks related to work have played a role, but it is also likely that health has increasingly been used as a criterion for the selection of the working population. Research evidence implies that the relatively less healthy have dropped out of the workforce because of redundancy, early retirement and work disability which resulted in a healthier working population in the 1980s.

The Work and Life Situation Surveys provide data on differences between the various industrial sectors in terms of quality of work. The most notable feature of this comparison was the relatively bad score of the transport industry in quite a number of respects (monotony, noise, dangerous work, irregular hours). After transport, the industrial sectors with the poorest scores are manufacturing and construction, which are more or less on a par with each other.

Experts are rather optimistic about the future of work and health. The aspect of working life which they feel on the whole negative about concerns stress-inducing factors (stressors), such as work pressure, workplace, shift-work and the absence of an adequate relationship between education level and job content. Thus, further policies have to be aimed at these aspects of the quality of working life.

Strengths and weaknesses of the occupational health policies

An evaluation study has shown that the Working Conditions Act had a stimulating influence on the improvement of working conditions, but that this was not a significant issue in many companies; many chief executives, for instance, have delegated the area of working conditions to subordinates. The OHSs have no policy influence in most companies and appear to play no significant role with regards working conditions.

Recent legislation on sickness absenteeism and work disability has been successful. The cluster of economic incentives introduced has had the effect of substantially decreasing sickness absenteeism and work disability.

To further elucidate the strengths and weakness of occupational health policies on employee health, the results of the Delphi Study of the TNO Work and Health Future Study are relevant. In that study it was concluded that both policy areas - the improvement of working conditions and the extension of occupational health care - have a favourable influence on all aspects of work and health. The influence, it was concluded, especially reduces the psychological stressors and the ergonomic and toxicological problems. It became clear that in both areas more active policy measures are being taken regarding work disability.

Priorities for action

Action is needed to improve the participation in the workforce of unemployed, sick and disabled people, as well as to improve the participation of women, elderly people, migrants, etc. Another priority for action is the problem of the serious mental workload and its effects on mental health.

Labour participation

The issue is how to keep sickness absenteeism and work disability within acceptable bounds, and also how to allow more people to participate in the labour process. In other words, how is it possible to get women, older people in less good health, and people partly disabled for work back into the labour force, and how can one stop the selection of employees on health grounds? Research must be done on optimal ergonomic adjustments, organisational changes, new patterns of working hours and so on, in order to expand the possibilities of employment for these groups. Increasing the number of day nurseries and provision of after-school care, for instance, is one of the most important measures which would allow more women to take part in the labour force. For older people, more part-time work, more part-time early retirement, a personnel policy that is more aware of the needs of different age categories, and vocational training especially directed to those returning to the labour process, are additional measures that will help increase their participation.

Mental workload and health

The second issue is that physical health problems are now slowly being replaced by mental health ones. The poor match between the level of education of workers and job content, the demanding working pace and heavy workload can all create mental problems and this is likely to continue to demand attention in the future.

In order to limit the negative effects of heavy workload, pressured work pace, mentally demanding work, shift work and so on, a number of approaches could be adopted, including:

- adaptation of work to the abilities and needs of the employees
- the social support of employees
- career management in relation to stress problems.

These kinds of measures, however, need to be accompanied by changes in management approach and a more intensive social policy.

Also important in enabling firms to deal with these issues is the involvement of employers and trade unions in the quality of working life. In their negotiations with employers, trade unions and their members need to focus less on wages and other financial aspects of working conditions, and more on the issues of work pressure, optimal labour participation and investment in the quality of work. Financial resources should therefore be made available to carry out programmes which will remove bad working conditions and limit the negative effects of the factors discussed above.

Conclusions

The Dutch labour force is heavily oriented to the services, transport and communication sectors (70% of all employed people). Whilst there is a tremendous increase in part-time employment, many middle-aged, older and female people are unemployed and seeking jobs.

The trend in the 1980s with regards health and safety legislation was for regulation; nowadays it is for deregulation. In social insurance legislation there is a shift of responsibility from government to industry; economic incentives have been introduced to bring absenteeism and work disability down.

Occupational health care now covers about 40% of the workforce. A significant further increase is expected because of legislation on working conditions. The main trends in this area are a decrease in dangerous and monotonous work, but an increase in work with mental stress.

The global trends in health indicators are a decrease or stabilisation in occupational accidents, absenteeism and work disability. It is believed that the selection of a healthier workforce is an important reason for this trend; this selection also makes labour participation relatively low. The main health problems are musculosketal disorders and psycho-social disorders.

The policy issues for the near future will be:

- labour participation in relation to absenteeism and work disability
- the mental workload, stress and well-being of employees
- the working conditions of young and female employees, with attention given to the double workload of employees with children
- working conditions in the transport sector.

Research

In the future more older workers and women will have to be employed in the labour force because there will be fewer young people available (due to demographic changes and due to the fact that they will remain longer in education) and because more women with children will want to join the labour force. This means that research must be done on optimal ergonomic adjustment, organisational changes, new patterns of working hours and so on, in order to expand the possibilities of employment for older workers and women.

Two explanations have been given for the relative improvement of the health of the workforce during the last decade: better working conditions and the selection of personnel on health grounds. It has not been possible to present empirical proof about which of the two has had the greater influence. Research on this issue seems to be extremely relevant because it would throw light on the extent to which 'better work' or 'better employees' have led to better health in the labour force.

It is a draw-back that little is known about the long-term effects of work on overall well-being and whether decline in health status is caused by work and/or ageing that is independent of work. Research in this area is required. Older workers report a worse health status than younger people. They mention more backache and complaints of the musculoskeletal system and also more fatigue and stress reactions. Moreover, they make more visits to the general doctor and use more medicines. Because older workers have a much higher rate of absenteeism and work disability, the attention of policy-makers, care providers and researchers has been concentrated on this age-group. If, however, there is an intention to minimise this 'drop-out' process amongst older workers, it seems a good idea also to pay more attention to the work situation of younger workers in order to anticipate the problem.

1 The context of national occupational health and safety policies in Portugal

Economic structure

Portugal currently has about 9.5 million inhabitants. There has been a decrease in the population in recent years due to a low birth-rate resulting in a progressive increase in the average age. The active population, which consisted of 2,518,100 men and 2,024,700 women (48%–49% of the total population) in 1993, has been stable for the last two years. The trend has been for employment to fall in the primary sector and increase in the tertiary sector.

There has been a progressive increase in the number of enterprises since 1985, reaching 150,972 in 1992. Less than 100 people are employed in 98.1% of enterprises; the remaining 1.9% employ 44.8% of the total working population. The so-called 'super-enterprises' (over 500 employees) are no more than 0.3% of the total, but employ 23.1% of the working population. Between 1985 and 1992 there was an increase in the proportion of people working for small and medium-sized enterprises. At the same time, the number of enterprises with over 100 employees fell and this fall was particularly significant amongst super-enterprises.

The labour market

The rate of unemployment in Portugal has been between 4% and 6%, but forecasts for 1994 have indicated a rate of 7.2%. Agriculture and industry are the most affected sectors. Since 1991, unemployment has been worst in the 25–44 age-group and amongst women of all age-groups, except for those over 54. There has been a decrease in the number of employees belonging to the younger age-groups (12–14 and 15–24) since 1988. This is, generally speaking, a positive trend which indicates the choice of education over precocious employment. Amongst older people there has been a drop in the number of employees in the 55–59 age-group, and a stabilisation in the 60–64 and over 65 age-groups.

Precarious employment (excluding the self-employed) has fallen from 52.2% of the working population in 1988 to 48.2% in 1991. There has also been a drop in the number of short-term contracts from 19% in 1988 to 16% in 1991.

The highest percentage of part-time work (involving less than 35 hours per week) is found in the primary sector. Women present higher figures than men. No data is available on moonlighting, but about 17% of employees are engaged in shift-work.

Job-seeking by women has shown a significant increase in the last few years. This trend should continue, given the pressures on family income, and also for cultural and social reasons related to equal opportunities and social integration.

The available data on migrant workers, both European and African, is very limited and relates only to enterprises with over 100 employees, so it is an incomplete measure of the extent of the presence of legal migrants and also fails to account for the numbers of illegal migrant workers.

Industrial relations

In 1991 in Portugal there were 406 trade unions, 44 labour alliances, 28 federations and 3 confederations; in the same year there were 409 employers' associations. There is no direct relationship between industrial action and the formal bargaining system; thus, in 1989–90 there were fewer strikes and more agreements, whilst in 1991–92 there were more strikes and more agreements, and in 1992 there were more agreements (at sectorial level), but most strikes affected one enterprise only. There has been an increase in the number of IRCT (Instruments of Collective Labour Regulations) during recent years: there were 357 in 1988, 377 in 1989, 368 in 1990 and 385 in 1991.

There are very few workers' committees, although they do have a presence in some enterprises, notably in construction. In this industry, due to the high casualty level, the collective agreement allows for the creation of health and safety committees.

Values and philosophy in relation to health and safety

On 19 October 1990, a Financial and Social Agreement was signed between the Portuguese Government, the General Workers' Union, the Portuguese Trade Confederation and the Portuguese Industry Confederation. This agreement included measures and proposals related to safety, hygiene and health in the workplace; it included provision for the future education of young people, vocational training and the training of active workers. The agreement contained measures dealing with collaboration between the occupational medical services and regular medical provision, the development of data and knowledge on occupational hazards and prevention techniques, and the need to create occupational hazard prevention structures. This involves controlling risks in the workplace and changing the behaviour and alertness of employees, as well as giving priority to collective prevention and to reparation and rehabilitation. An Agreement on Work Safety, Hygiene and Health (1991) has also been reached between the Government and all the social partners. Its main objectives are to improve labour conditions and social protection. There are provisions aimed at the stimulation of creativity and motivation, the improvement of physical and psychological well-being, the lessening of individual, family and group tensions, and the reduction in occupational injuries and diseases.

The European dimension

The replies to the European Health and Safety at Work questionnaire show that 60% of Portuguese workers have a positive attitude to the European Union, 4% are negative about it, while 24% do not know and 12% ignore the subject.

2 Occupational health and safety policies and structures

Legislation

There is a wealth of Portuguese legislation on hygiene and safety at work; the first measures date back to the beginning of this century. Despite difficulties encountered when trying to implement many of the provisions, protecting workers' interests has always been a concern for the legislators. This concern led the Portuguese Government to create legislation for occupational medical services. These were complementary to the so-called environmental standards, in that they recognised the need for the periodic surveillance of those workers vulnerable to the more hazardous environmental exposures, and the need to study the best way to adapt such workers to their working conditions in order to improve both their personal satisfaction and productivity.

Occupational health and safety structures

Until January 1994, any company with over 200 employees or those whose activities were considered hazardous to human health, had to provide an occupational health service. Decree-Law No. 26/94 of 1 February 1994 stipulates that the organisation of safety, hygiene and occupational health structures is the responsibility of the employer and must cover all employees. These services can be organised on an internal, inter-company or external basis. Health activities can be organised separately from safety and hygiene activities.

Control and inspection

Under Decree-Law No. 219/93 of 16 June 1993, the General Inspectorate of Labour (GIL), which is a central service of the Institute for the Development and Inspection of Working Conditions, is in charge of inspecting and supervising compliance with standards related to working conditions, employment and unemployment. The GIL is responsible for the whole of the country and performs its duties in all branches of activity, in all public, private or co-operative enterprises, whether they have employees or not.

Information collected in 1992 showed that the GIL covered an estimated 143,000 enterprises, with 168,000 establishments, the most significant activities being in the retail trade, the building trade and public works, restaurants and hotels, and wholesale. A total of 98,517 inspection visits were made to 82,019 establishments in 1992.

There are 298 inspectors on active duty, 72 of whom have a degree and the remainder a secondary school diploma or equivalent. There is approximately one inspector for every 11,000 workers. The role of labour inspectors includes: visiting places of work; questioning the employer, management and workers when this appears necessary; ascertaining conditions pertaining to health, safety and well-being in places of work. Besides these functions, inspectors should also draw up reports of inquiries into accidents at work and occupational illnesses, take part in official inspections and monitor the company's obligatory duties to maintain and operate occupational medical services, as well as those of hygiene and safety at work.

In 1992 there were a total of 1,566 lawsuit reports on infringements of rules, 135 of which related to occupational health, safety and hygiene, and 20,332 lawsuit reports on infringements of contracts, of which 4,909 concerned occupational health, safety and hygiene. In the same year, 14,253 establishments were visited for safety, hygiene and health at work purposes, and 6,636 technical reports were made. This resulted in 5,176 lawsuit reports of which 40.9% were related to accident reports, 16.0% to poor safety and hygiene conditions and 32.1% to labour insurance. These three areas constituted nearly 89% of the total number of violations detected in the field of safety, hygiene and health at work during 1992.

Occupational health services

Occupational health services (OHS) are required to supervise the capability of employees to carry out their duties and to oversee their health. These services, either by themselves or in collaboration with other specialist services within the same company, must study and supervise, in particular:

- the company's hygiene and health conditions
- the collective and individual protection of the employees against fumes, gases, vapours, dust, noise, vibrations, ionizing radiation, accidents and occupational diseases
- ergonomic considerations
- the response of the employees to the various services.

Only 12.9% of the active working population, or 606,113 employees, are supported by OHSs.

Doctors working in OHSs belong to three categories: they may be specialists in occupational health, doctors who have completed a course in occupational health in one of Portugal's three main university centres or doctors with no specific training who hold a temporary permit from the Ministry of Health.

Employers are responsible for the financial support of their OHS. The doctors attached to these services retain their technical-professional independence, although they depend, both financially and administratively, on their employers. Occupational health doctors are directed and supervised by the Directorate General of Health, and enterprises are answerable to the latter if the regulations laid down in Decree No. 47512 are not complied with. But otherwise there is virtually no relationship between OHS and the general health system.

Information and assistance

The former Directorate General of Hygiene and Safety at Work used to publish a monthly bulletin aimed at workers - *Boletim de Prevenção no Trabalho* - which will be restarted; it had a print run approaching 70,000 copies. The Portuguese Society of Insurers publishes a quarterly magazine, *Segurança*, with a subscription circulation of 3,000. Additionally, various brochures, publications and monographs aimed at the working public and officials working in the areas of safety, hygiene and health at work have occasionally been published by different concerns.

Training

According to inquiries made by the EC during the European Year of Health and Safety at Work, 84% of workers had not attended any training courses in safety, hygiene and health at work. It is therefore perhaps not surprising that 86% of the workers interviewed during that survey expressed an interest in receiving more information about health and safety protection. In 1990-91, nearly 500,000 people attended vocational and technical training courses; 57% of these were actively employed, but many of these courses contain no health and safety components.

Medical experts on occupational health receive training first at the National School of Public Health and later at the faculties of medicine in the Universities of Oporto and Coimbra. A Master's course on Occupational Health was created in 1981 in the Faculty of Medicine of Coimbra. The Ministry of Education is responsible for courses on hygiene and safety at work in various trade schools (Aveiro, Barreiro, Estarreja, Avelar, Lisbon and Oporto).

These are open to pupils who have completed at least their 9th year at school. The inclusion of safety, hygiene and health at work in the school curriculum is currently under consideration.

Research

Occupational health research is undertaken by both public and private bodies, notably the Universities of Coimbra, Oporto, Lisbon, Aveiro, Minho, the National Public Health School, the National Health Institute, the Institute of Welding and Quality, the General Aeronautic Material Workshops at Alverca, and the Postal and Telecommunications Services of Portugal. There is no agreed policy on research. Each institution plans and undertakes its own research activities within the area of safety, hygiene and health at work. Staff involved include university professors, doctors, engineers, hygienists, psychologists, sociologists and lawyers. There has been a recent tendency to increase research in the field of occupational health as a result of the growing numbers of specialised doctors and the increased interest of universities and private institutions. Fields covered include, noise, lighting, oncology, epidemiology, stress, infections, ergonomics, computers, alcoholism, dermatoses, ageing, epilepsy, asbestos, ophthalmology, health conditions and hazard factors in various activities, solvents, heavy metals (lead), occupational deafness, chemical hazards, accidents at work, anthropometry, the evaluation of hearing protectors, pneumoconiosis, industrial pollution control, thermal environments, vibrations, promotion of health in the workplace, safety, drug abuse, dust and the organisation of OHS.

Occupational health insurance

Insurance is applied by law to all employees and is considered a commercial activity. Civil servants, the armed forces and certain major, public enterprises maintain their own insurance schemes and are excluded from the main scheme. The aim is that insurance provides victims of occupational accidents and those suffering ill-health with adequate assistance in the form of medical care, rehabilitation and income during the period of illness, as well as compensation in cases of permanent disability. Compensation is the scheme's main activity. Insurance premium rates vary with the number of accidents and the hazards identified, so that in those enterprises where there are fewer accidents and better working conditions the premiums are lower.

3 Health and safety outputs

Occupational accidents

Table 6 shows overall figures for injuries and fatalities from occupational accidents between 1988 and 1992.

	1988	1989	1990	1991	1992
Occupational accidents	290,961	304,636	305,512	293,886	278,45 5
Fatal accidents	595	288	203	224	185

Table 6: Occupational accidents - evolution from 1988 to 1992

Occupational diseases

Replies to the question 'how does your work affect your health?' in the European Year of Health and Safety at Work questionnaire, suggested that it affected health in the following ways:

27% - eyesight

28% - stress

20% - hearing

55% - aches and pains

58% - back pain

22% - breathing problems

51% - general fatigue.

According to statistics collected and published by the labour inspectorate, pneumoconiosis accounts for about 43% of recognised occupational diseases.

Absenteeism

According to the analyses published by the Ministry of Social Security, about 7.5% of absenteeim is caused as a result of accidents at work and 57.2% is the result of illness, although this includes non-occupational illnesses.

Work environment surveys

Many studies show that occupational noise is a serious hazard, as is exposure to organic solvents, heavy metals and thermal stress. Pneumoconiosis, dermatoses, excessive workload, stress and the causes of accidents have been the main subjects of study on the part of Portuguese researchers.

4 Assessment of occupational health and safety policies

The main problems related to health and safety are:

- limited coverage of employees and enterprises by occupational health services
- the lack of analysis and the poor credibility of statistics
- inadequate co-operation between the various bodies responsible for these matters
- the poor quality of services rendered to small and medium-sized enterprises and independent workers
- the adaptation of enterprises to new regulations
- difficulties in implementing an efficient prevention technique
- the absence of a prevention culture both at employer and employee level due to the lack of integration of health and safety in the educational system
- insufficient applied research
- a low level of understanding and acceptance in relation to the need for investment in occupational health and safety.

The strengths of the present system have been identified as:

- the support it provides for employees who become ill
- the potential for controlling disease, including occupational diseases, under current legislative requirements, which is enhanced by compulsory medical examinations for occupational diseases and disabilities.

Furthermore, provided that the contribution of occupational medicine and general medicine are properly co-ordinated, the development of OHSs and greater openness of the medical profession when dealing with workers should be positive factors in combatting the problem of absenteeism resulting from occupational illness and injury.

The weak points of the present system reveal its lack of strategy which is the consequence of a weak occupational health policy. This is illustrated by the Government's creation of the Institute for the Development and Inspection of Working Conditions not only some 18 months after the deadline it had undertaken to meet, but also with a character and function very different from that defined in the Agreement on Safety, Hygiene and Health at Work in 1991. This Agreement stipulated the creation of a tripartite institute with the flexibility and capacity to lead in the field of health and safety. What the Government has actually created is an organisation that merges the new tripartite institution envisaged by the Agreement with the Labour Inspectorate and it has given the resulting institution additional responsibility for all aspects of industrial relations.

The European Union will continue to affect health policies, but this will not necessarily cause problems, it may even be an advantage. However, an overall activity scheme with specified objectives in which EU Directives could be transposed according to their potential for implementation is required.

5 Conclusions

There is a growing concern in Portugal about safety, hygiene and health at work. Concern is shown by the authorities and the social partners, as well as by universities and other institutions. Difficulties have been encountered in implementing several of the measures proposed in the legislation and agreed upon by various employers. Economic, social and cultural factors are partly responsible, such as the lack of a 'prevention culture'. The general outlook, however, is optimistic and it is believed that there will be better understanding in the medium-term of the importance of this issue at both national and workplace levels.

1 The context of national occupational health and safety policies in Spain

Economic structure

The economic and employment context in Spain has changed substantially over the past decade. During this period the service sector has grown very rapidly and the workforce in industry and agriculture has decreased, whilst the construction sector has been subject to oscillations related to variations in the general economic cycles.

In 1993 over 7 million people were employed in the service sector, 2.56 million people in industry and 1.1 million in agriculture.

The labour market

Following a tendency to increase from 1991 to 1994, Spain's unemployment has being decreasing in 1994-95, from 17.3% at the beginning of 1994 to 15.3% at the end of 1995. Unemployment is higher in the areas affected by the industrial recession - textiles in Catalonia, footwear in Alicante and steel foundries in Asturias. On the regional level, the highest unemployment is found in southern Spain, and the lowest rates are to be found in northern Spain: Navarra and Rioja. The unemployment rate of women and young people is above the average (26,3 and 36,7% respectively).

Over the past 10 years there has been a significant immigration of both non-EU citizens as well as those from EU Member States into Spain. Especially important has been the entry of people from the Magreb and parts of West Africa.

Since 1981 there has been a steady increase in the employment of women. In absolute terms, in 1985 only about four million women were economically active, while by 1993 this figure was over 5.5 million.

Industrial relations

Union membership in Spain is about 8% of the employed population. Nevertheless, all workers have the right to elect delegates and representatives to the workers' committee. The two major trade unions are Comisiones Obreras (CC OO - Workers Commissions) and the Union General de Trabajadores (UGT - General Union of Workers). Although there are no official statistics available on the subject, surveys indicate that about 40% of firms have active contacts with employers' organisations.

Collective bargaining is carried out at several different levels. Especially important are the sectorial agreements at national level which are complemented by company- and regional-level agreements.

2 Occupational health and safety policies and structures

Legislation

Article 40.2 of the Spanish Constitution (1978) -under the sub-heading of Chapter III 'On the Guiding Principles of Economic and Social Policy'-, states that the public authorities will establish measures for safety and hygiene at work. Article 43 mentions the role of the authorities in the promotion of health education.

Law 31/1995, 8 November 1995, of Prevention of Occupational Risks has replaced the General Ordinance on Health and Safety at Work 1971, which covered people within the social security system, and has introduced a broader general objective, to improve occupational health and welfare, extending prevention to all persons at work in all types of occupation.

The Law provides a basis whereby employers and workers themselves can solve their working environment problems by means of collective bargaining and planning prevention of occupational risks in the workplace or enterprise. It makes provisions for health and safety delegates and joint health and safety committees. Health and safety delegates - elected by and from among the workers' representatives in an enterprise or workplace - have monitoring and advisory duties as regards prevention of occupational risks at work, in collaboration with the Directorate.

The Health and Safety Committee is a joint body responsible for preparing, putting into practice and assessing the plans and programmes for risk prevention. These committees consist of technical specialists, workers' representatives and persons appointed by the enterprise. The are compulsory in enterprises or workplaces with more than 50 employees and in those where particularly dangerous activities are carried on.

The provision of information in many forms, and the promotion of education and training in safety and health plays a large part in achieving the improvement of health and safety at work. The Law sets up a framework from which further regulations will implement the more technical aspects of risks prevention. The Law implements the Framework Directive (89/391/CEC) as well as Directives 92/85/CEC on pregnancey/protection of maternity liave; 94/33/CEC on occupational health and safety in limited duration employment and temporary employment enterprises and 91/383/CEC on temporary and fixed-term employees.

Another fundamental text which deals with the participatory rights of workers and their representatives is Article 19 of the Workers' Statute Law (1980). Also relevant is the General Law on Health (1986) which defines the functions of health systems in Spain, including occupational health. In addition the Decree revising the General Law on Social Security (1974, amended 1994) introduces some relevant points concerning the prevention of risks at work.

Occupational health and safety structures

The Ministry responsible for devising and implementing occupational health and safety policy is the Ministry of Labour and Social Security, although the Ministries of Health and Industry have responsibilities in this field. The management of occupational health and safety services is the responsibility of the Autonomous Communities, whilst the Ministries of Labour and Social Security, Health and Industry are responsible for regulatory tasks, planning policy and coordinating the various bodies.

Control and inspection

In 1994, there were 1,448 civil servants involved in inspection according to Ministry of Labour statistics (the figure includes 616 inspectors and 832 controllers). The Spanish Labour Inspectorate undertakes the enforcement of laws on employment, social security, health and safety at work, trade union's law and individual and collective rights of workers. The area of health and safety is solely within the competence of inspectors, who also assume responsibility for the rest of the areas, whilst controllers act only on employment and social security issues in firms of less than 25 workers.

Labour inspectors are university graduates who are selected through a state examination system involving rigorous testing in labour law, labour relations, social security and occupational health and safety, as well as general subjects (economics, law, sociology and accounting).

There were just 109,912 visits to workplaces related to health and safety with 16,208 involving fines (in relation to 33,314 law infractions) and 65,166 requirements and 796 orders to stop works or tasks in 1994.

Occupational health insurance

All wage earners are covered by mandatory occupational health insurance which is an integral part of social security. Most employers have contractual arrangements with the Mutuas de Accidentes de Trabajo y Enfermedades Profesionales (Mutual Societies). These are associations of companies that are privately managed and funded by insurance premiums from employers, although they are strictly controlled by the social security system. Recent figures indicate that 9,120,543 workers are covered in this way, which represents a 9.09% increase over the previous three-year period. An additional 1,458,339 workers were covered by groups associated with the mutual societies in 1991. The rest of the workforce is covered by the social security system, which acts as their mutual society. The mutual societies are responsible for 760 medical/assistance centres throughout Spain. These are staffed by 11,926 health and administrative professionals.

Occupational health services

Company medical services cover 15% of Spain's workforce. They are obligatory in firms with more than 1,000 workers. In the case of enterprises with more than 100 workers, autonomous health services can be created in collaboration with similar firms. The Ministry of Labour can oblige firms of less than 100 workers to set up their own medical services in certain situations, such as where there are high occupational risks for workers, as in the handling of heavy metals, for example.

Information and assistance

Information campaigns in relation to health promotion and workplace environments have increased both in quality and quantity during the past 10 years. Some autonomous regional governments have carried out information campaigns in collaboration with the Labour Inspectorate; these are aimed at reducing accidents in certain industries, notably in construction.

Training and research

Training is now becoming an important issue for trade unions and a variety of levels of courses are provided for rank-and-file members as well as trade union representatives and officials. The task of training is shared between the trade unions and the National Institute for Health and Safety at Work (INSHT). Since 1987, they have signed a series of collaborative agreements on information materials and courses on occupational health and safety. INSHT gives courses at three levels:

- basic courses for the workforce aimed at raising awareness,
- intermediate-level courses for active trade union members and officials, middle managers and technical school teachers,
- specialised seminars and more technically orientated courses for professionals in the field of occupational health and safety.

Trade unions also organise their own courses for their members and for trade union delegates and officers.

Research is largely located in the universities, schools and institutes of public health, specialised institutions (such as INSHT) and within the central and autonomous administrations.

3 Health and safety outputs

Occupational accidents

In the decade 1981–1990, 11,852 workers died in Spain as a result of occupational accidents. However, it should also be pointed out that all non-traumatic deaths are included in the mortality data and that a large proportion of fatal accidents are in fact traffic accidents that have occured to the victim up to one hour before the start of the working day and one hour after its end. 60% of deaths were concentrated in just a few sectors of economic activity involving less than a quarter of the working population. There has been a more or less steady increase in the death rate from occupational accidents during the decade, with greater increases in certain areas from 1987 onwards.

Over the same ten-year period, 117,832 serious accidents were registered. In absolute terms as well as in rates, serious accidents reflect the occupational mortality pattern. There has been a steady growth in serious accidents with a peak in 1988.

The traditional view of the causes of accidents at work is that they are a result of a combination of what can broadly be considered environmental variables and the 'human factor'. In recent years this approach has been criticised for being both partial and for 'victim-blaming'. A more up-to-date causal model of occupational accidents and ill-health at work takes into account such factors as, among other things, how work is organised, working conditions, contractual conditions and labour relations.

Occupational diseases

Changes in the system of production, the ways of working and working conditions have resulted in significant changes in occupational diseases in Spain, with a move away from classical pathologies to non-specific pathologies. This makes linking working conditions with disease increasingly difficult.

However, the majority of declared occupational diseases are contact dermatoses, diseases produced by physical factors, occupational deafness, certain infectious diseases such as brucelosis and hepatitis, respiratory diseases and chemical poisoning.

Absenteeism

There has been an increase in sickness absence over the last few years and, consequently, an increment in the average number of days lost. In economic terms this means that 13.79% of the expenditure of mutual societies covers sickness absence.

4 Assessment of occupational health and safety policies

The main actors in occupational health and safety - the social partners, health professionals and the public administrations - have suggestions for policy changes. While there are some differences of detail amongst the trade unions, they would like the new Law of Prevention of Occupational Risks to be implemented. The employers are, in general, concerned about 'over-regulation'. Health professionals and trade unions express their concern about the inadequate classification and registration of occupational diseases and call for the imposition of an updated health information system.

There was considerable divergence in the views of the social partners on the development of occupational health and safety policy and strategies in Spain. Trade union representatives interviewed base their approach on a demand for more strenuous enforcement, but indicate that in an economy with such a high level of unemployment and continued threat of factory and plant closures, the acceptance of lower health and safety standards is a likely outcome. The unions also point out that increased flexibility in the labour market is leading to increases in occupational accidents (for example, in construction and agriculture). They regard the EU as a beneficial Agency for Spain and seek some kind of convergence with the standards they perceive to exist in other member states.

The employers, on the other hand, are concerned about further regulation, arguing that this will affect their ability to be competitive in both national and international markets. They would prefer to see more self-regulation within industry, rather than the imposition of further punitive measures. Flexibility is strongly supported by employers who argue that it allows for job creation and they do not see it as a cause of increases in occupational accidents.

It appears from the interviews with representatives of the social partners and health and safety professionals that there are many issues on the policy agenda as far as occupational health and safety are concerned. It seems that there is a need to take into account the health promotion paradigm in the workplace as well as the issue of working conditions and the working environment. Current Spanish legislation on occupational health and safety is now in line with EU provisions and is likely to be implemented following these criteria.

Apart from the need for the relevant administrations to make greater efforts to co-ordinate their activities, it is also imperative that the insurance organisations shift the focus of their work towards increasing the incentive for preventive health and safety in enterprises. At the same time, the Labour Inspectorate should be provided with more human and economic resources so that it can comply with its brief. If this is not done, new legislation will not be enforced because of the lack of the necessary infrastructure. Also relevant to the resource issue is the creation of advisory bodies and the support for research which will lead to the establishment of more effective preventive practices in Spain.

Both work-site medical services and primary health care need to place greater emphasis on prevention and health promotion in the workplace. Co-ordination of their activities should be undertaken in order to move towards the unification of their health information systems.

The changing nature of work and work-related problems requires the training of new professionals in the field. New university studies in labour relations, health education and promotion, public health and communication sciences are required to provide a new set of technically competent professionals able to deal with the rapidly changing world of work and health.

Health and safety committees in the workplace have an important role in prevention and health promotion at work. However, for worker participation to be effective in this context, it is necessary to secure workers the right of access to information in relation to the processes of production. Training in occupational health and safety matters may also serve to stimulate worker participation in health issues.

The formation of teams of workers trained in occupational health within the trade unions themselves is necessary to help to overcome the very serious problems of working conditions and environments. Public administration could provide technical support for such teams.

The trade unions have an important role in changing worker perception about health and safety risks at work. Trade unions have made efforts to move away from dubious practices, such as 'danger money' (pluses de peligrosidad), and are concerned about guaranteeing adequate working environments through their own education and training programmes.

Most occupational health professionals argue that the welfare of workers has implications for employers given that better health status implies less absenteeism, greater productivity and a reduction in costs to the firm. The involvement of management in occupational health and safety is therefore as important as worker participation. But for this approach to be effective, employers will need to accept that improved health standards is a factor in the economic success of the enterprise.

Concrete actions to improve occupational health and safety should be evaluated and the successful ones publicised. Evaluation is a basic instrument for achieving the goals of prevention and health promotion at work.

1 The context of national occupational health and safety policies in Sweden

Economic structure

Historically an agricultural country, the proportion of people employed in agriculture in Sweden has fallen dramatically from about 78% in 1855 to 3% in 1990. Initially this was caused by industrial growth. The demand for iron and timber products, the building of the national railway system and the mechanisation of agriculture caused a rapid growth in industry, trade and communications. Since the 1950s, however, Swedish industry has stagnated and now employs only about 26% of the workforce. The growth has been in the service sector, now employing 70% of the workforce.

The number of small enterprises with less than 50 employees has increased over the years and about 50% of employees are found in such enterprises. From 1966 to 1985 women entered the labour market in increasing numbers; education, expanding child care amenities and the rising cost of living are some of the factors accounting for their increasing participation. In 1992 the workforce comprised 83% of men and 79% of women, but about 40% of the women worked less than the normal 40-hour working week. More men (63%) than women (37%) were employed in the private sector, while more women (70%) than men (30%) were employed in the public sector.

The labour market

The level of unemployment in Sweden remained low for a very long time until recently, but with the growing economic crisis it rose from 1.5% in 1989 to 5.3% in 1992. In 1993 it exceeded 8% for the 16–64 age-group. Unemployment in the younger age-group (16–24) is about three times as high (18.5%) as in older age-groups (25–64); it is also somewhat higher amongst men than women, and particularly high amongst immigrants (20.8%). Furthermore, almost 6% of the workforce is in government-sponsored labour market programmes.

Industrial relations

Sweden has had a system of industrial relations dating back to the Basic Agreement of 1938 between the Swedish Employers Confederation (SAF) and the Swedish Trade Union Confederation (LO). A number of further agreements have been signed in the ensuing years. During the 1970s, industrial relation issues that were previously regulated by agreements came to a great extent to be regulated by legislation. More recently, several national agreements have been abandoned by SAF, such as the agreement on occupational health services and the collective bargaining system. SAF has also left all the Boards of the state tripartite decision-making structures, causing the Government to ask the unions to do the same. This has meant that the Swedish model of tripartite co-operation between the state, labour and management has lost some of its importance. The problem stems from a divergence of approach, with the LO preferring to retain centralised bargaining, whilst the employers are seeking a more decentralised arrangement. In this situation the future of pay bargaining remains uncertain and the issue continues to be debated. However, local agreements remain in operation and most Swedish enterprises still have good industrial relations in the workplace.

The social partners have collaborated on industrial safety over a long period. They became aware at an early stage that it was in their common interest to combat accidents and hygiene hazards. Hence, safety committees and safety delegates are well established institutions in Sweden. All workplaces with more than 50 employees are required to have a safety committee and almost all workplaces with at least 5 employees have safety delegates. The safety delegates are, if possible, appointed by the local trade union organisation which currently or traditionally has a collective agreement with the employer. For some years, the number of union-organised employees decreased, but the number is now rising and 80–85 % of the workforce are members of trade unions.

The European dimension

The basic position of the Swedish Government is that Sweden wholeheartedly supports the fundamental goals of the European Union as formulated in the Treaty of Rome and the Maastricht Treaty. In a number of specific areas, however, Sweden is negotiating special, mainly temporary, exemptions, such as in regional development policy, agriculture, environment, and health and safety, where Sweden has more far-reaching policies than most EU countries do. Before joining the EU, polls indicated that many Swedes believe that the advantages of membership of the EU will outweigh the drawbacks. Today a majority of Swedes would not vote for membership because of disappointment that the promised reduction in the price of food and in interest rates have not yet materialised.

2 Occupational health and safety policies and structures

Legislation

In Sweden the concept of the work environment is defined broadly. It not only refers to the traditional physical and technical environment, but also to the way work is organised, job content, working hours, opportunities for personal development, co-determination and the psycho-social aspects of the workplace.

The current Work Environment Act came into force in July 1978 and increased the rights of unions to help improve the work environment. The Act contains basic provisions concerning occupational safety and health issues in Sweden. More specifically, it includes rules on how employers and employees should co-operate on work environment matters. The main responsibility for the work environment is borne by the employer.

The Act stipulates general requirements which apply to the physical and psychological work environment. The purpose of the law is to guarantee a working environment that does not expose employees to ill-health or accidents, and that is satisfactory with regard to the nature of the work and the social and technical development of the community. The law is also designed to promote co-operation between employers and employees to achieve a good working environment.

The Act has been subject to several amendments. An additional new provision requires the employer to ensure that appropriate job modification and rehabilitation activities are conducted at the workplace. A further new amendment concerning internal monitoring requires employers to know the extent to which their enterprise complies with the relevant occupational safety and health regulations and to draw up a timetable as to when any measures necessary for compliance will be taken.

Recent amendments to the Act have been introduced to facilitate harmonisation with European Union law in line with the implementation of the European Economic Area Treaty. Provisions in the Swedish work environment sector will not be radically changed by the adjustment to the EU, but intensive work has been devoted to the implementation of EU Directives, involving systematic comparison between the rules of the National Board of Occupational Safety and Health and all the EU Directives relating to the work environment.

The Work Environment Act applies to every activity in which employees are used for work on an employers' behalf. The definition of employee also includes persons undergoing training or education, inmates of an institution who perform work they have been allotted, and conscripts or other persons performing statutory service or participating in voluntary training for activities within the national defence establishment.

Occupational health and safety structures

Work environment issues are under the purview of the Ministry of Labour. The National Board of Occupational Safety and Health is reponsible for the central supervision of compliance with work environment and working hours legislation.

Control and inspection

The main task of the Labour Inspectorate is to ensure that enterprises and administrations comply with work environment legislation in order to prevent work-related injury or illness. The range of activities carried out by the inspectors include:

- inspections
- counselling and information
- advance assessments of plans of working premises, personnel facilities, work processes and working methods

- work injury investigations.

Projects are conducted in areas of special priority or by focusing on high-risk groups in certain industries. Inspectors concentrate on workplaces that need numerous inspections, basing their priorities on knowledge of the hazards in different economic sectors and occupations, as well as conditions at particular workplaces. They are required to take an holistic view in their supervisory activities. Each year, from Sweden's 240,000 workplaces employing a workforce of more than 4 million people, some 400 inspectors visit roughly 38,000 establishments.

Occupational health services

The occupational health service (OHS) system in Sweden grew very quickly up to 1992 as a result of an agreement between the social partners and with the help of grants from the state. In January 1993 the agreement was cancelled and the grants have been withdrawn, the Government reasoning that since work environment matters are the responsibility of employers, they should also be responsible for preventive measures. The services are now totally financed by the clients. This has caused financial problems for some OHS units which have had to reformulate their services and reduce personnel. It is possible that the OHS organisation grew too quickly and now it has to find the right balance between its services and the clients' demands. There are no official statistics on the extent of OHS in Sweden, but an estimate is that about 60% of the Swedish workforce is covered at present. In some sectors the coverage is known to be greater; 85% of the employees of the Swedish Association of Local Authorities, for example, were covered in 1993.

The occupational health services are often organised into teams of occupational health doctors and nurses, safety engineers, physiotherapists and behavioral scientists. Their work covers the medical, technical and psycho-social aspects of the working environment and is essentially preventive in nature. Large companies often have their own internal health service. Small and medium-sized enterprises use joint company health centres serving a number of enterprises in the same geographical area. There are also some industry-specific units whose work is directed at enterprises within the same industrial sector.

Research

Swedish occupational safety and health research is well-known internationally in such fields as musculoskeletal injuries, ergonomics, occupational cancers, chemical health hazards, stress and systems of work. Research into workplace democracy, job content, personal development and the introduction of new technology and new production systems are also important areas. The largest contributor to such research was the Work Environment Fund; this was financed by a small work environment levy which was included in the social security contributions paid by employers. The levy brought in more than 55 million ECU each year.

The Work Environment Fund operated in three main areas with the aim to improve conditions at the workplace:

- research, both applied and basic
- development, including local development work at workplaces and in the regions
- dissemination of knowledge and working environment education - for example, research and development findings are presented in the Fund's own Swedish and English-language publications.

Many projects financed by the Fund are still in progress although the Fund itself has closed down. The role of the Fund has been taken over by the Swedish Council for Work Life Research, whose activities are not yet finalised.

Sweden has a number of research establishments which serve the needs of specific industries or areas of activity. The two largest, the National Institute of Occupational Health which specialised in more traditional work environment research, and the Swedish Institute for Work Life Research which dealt with labour market issues, work organisation and industrial democracy have recently merged into one research institute on working life.

Occupational health insurance

Work injury insurance is the oldest form of social insurance in Sweden and dates from 1901 when the first legislation on the subject was passed. Since 1978, under the Work Injury Insurance Act, all economically active persons - employees, employers and self-employed persons - regardless of nationality have been compulsorily insured against work injuries. Self-employed people must be domiciled in Sweden. The law does not make exception for members of family businesses. People undergoing training are also insured for work injuries in so far as their training involves such risks. Accidents occurring on route to or from work also qualify as work injuries.

In 1992 the Work Injury Insurance Fund showed an accumulated deficit of more than 2.9 billion ECU. Consequently, since 1993, the evidence requirements for work injuries have tightened in order to lower the number of cases accepted. In 1992 a new sick-pay system was introduced which shifted the financial burden from the public sector to the employer during the two first weeks of illness. There is no compensation for the first day of absence.

Also since 1993, the work injury insurance scheme has followed the same cash-benefit system as that of social insurance in general, with a large number of the total work injuries reported being settled during the first 180 days as ordinary sickness benefit cases. In cases leading to more than 180 days of sickness absence, the Central Office of the Social Insurance Service decides whether or not the injury is due to conditions at work. There is no list of approved occupational illnesses.

Monitoring

The Swedish Occupational Injury Information System (ISA) is a database on occupational injuries and job-related illnesses which is maintaind by the National Board of Occupational Safety and Health. ISA was designed for compiling official statistics and performing special analyses requested by various clients. ISA also presents an annual report - *Occupational Diseases and Accidents* - in collaboration with Statistics Sweden. The labour inspectors have local systems (SARA) with detailed facts on occupational accidents and diseases at each workplace to help them to plan visits and supervision.

3 Health and safety outputs

Occupational accidents

Statistics on work accidents have been kept since 1906 and statistics on cases of occupational diseases since 1936. The frequency of fatal accidents fell by 85% between 1955 and 1990. It has now increased again somewhat, but work accident mortality in Sweden is low by international standards.

Accidents resulting from musculoskeletal strain are the commonest type and there was a conspicuous rise in such accidents until 1987, but the trend has moderated in recent years. Some occupations are more injury-prone than others. The results of violence or threats of violence at work have become more common and three out of five reports concern women.

The risk of work accidents is greatest among the youngest employees (16–24), although the extensive efforts made by the Labour Inspectorate to reduce accident rates among young employees are now yielding good results. The economic recession is another factor affecting accident rates for young persons.

Occupational diseases

Reports of work-related diseases almost tripled between the beginning of 1980 and 1988; since then there has been a significant reduction. The risk of work-related diseases increases with age.

Work environment surveys

Statistics Sweden (SCB) undertakes continous surveys of working conditions in Swedish working life by questioning between 10,000 and 20,000 people every other year.

Concern during the early 1980s about rising trends in the number of work injuries, early retirement and disability pensions caused the Government to appoint a Commission on Working Conditions in 1988. This resulted in a report on a number of occupational categories that were subject to particularly serious working environment problems, including manufacturing occupations with severe and repeated accidents, hearing losses and musculoskeletal injuries. The Commission was critical of the effectiveness of occupational safety and health efforts at the workplaces it investigated, and also found that information systems for monitoring needed to be revised.

To increase funding for working environment improvements and rehabilitation programmes, the Swedish Working Life Fund was set up in 1990 by a decree from the Swedish parliament. The money was collected from employers in the form of a special charge and was originally to have lasted for six years; however, it was closed after five years. An evaluation of the effects of the grants given for projects to reform work systems (49%), improve the physical environment (24%), rehabilitation (21%) and new technology (3%) is under way.

4 Assessment of occupational health and safety policies

The State

A Labour Inspectorate with three inspectors was originally initiated in 1890 for the protection of the life and health of workers at work. In 1993 about 1,700 people were working for the Inspectorate in a variety of tasks in the work environment area, including supervision, financing, research, education and information.

During the last few years there have been changes in policies on occupational health and safety strategies brought about by the need to cope with a very serious recession and by the economic and politial policy, first of a Conservative Government and then by a Social Democratic Government. Further changes are due to a review of the way the Labour Inspectorate conduct their supervision, the influence of European Union membership, as well as new work environment problems.

The Government has started to make cuts in public expenditure in a number of different areas. Prompted by the cost of work injury insurance exceeding its funding by billions, there has been an extensive review of the social insurance system leading to changes in both the concept of work injury insurance and the compensation system itself. The Government has also closed down the Working Life Fund which was set up to reverse the trend of long-term sick leave and early retirement.

Today Sweden is experiencing the highest unemployment level since the 1930s, with the result that the prospect of unemployment has added significant strains to the current situation.

Although it is generally claimed that the standards of occupational health and safety will not be lowered, it is likely that the recession will have a negative impact on willingness to invest in the work environment. The economic situation means less resources for health and safety and most of the participants in work environment matters in Sweden voice concern about the recession and the drive to cut expenses. Some have expressed a strong feeling of uncertainity about the future of Swedish work environment policies and anticipate further changes in priorities. They all argue that a healthy work environment is a necessity and that work towards such a goal must not stop. It is recognised, however, that it may now take longer to accomplish. There is a very strong will to work for a better work environment, but there are problems concerning the relationship between some of the disciplines - such as research, implementation, supervision and training - which have their own policies but do not always relate to each other as they should.

The employers

The employers' organisations argue that health and safety should be integrated into the regular policy of the organisation. They point out that preventive work for safety and health is a must and that health and safety are prerequisites for efficient production, but that issues of comfort may have to come second. Their goal is to provide a good working environment in all its aspects. The employers, however, question whether the enormous investment in health and safety up to now has been sufficiently effective.

The employers'organisations, therefore, want their members to integrate health and safety issues into the general policy of their enterprises. This is already required in the internal control provisions of the amended Work Environment Act, which state that the employer is responsible for systematically planning, directing and inspecting the safety of the working environment (internal inspection), as well as continuously investigating hazards and work injuries. The employers' organisations are trying to reach all their members with this message about the need for quality assurance and internal control as well as the importance of co-operation in the workplace. They admit, however, that there are difficulties, particularly in reaching small enterprises, that require special attention.

The trade unions

The unions are proud of their continuous record on work environment matters and stress the importance of recognising new problems in the work environment, for example in new production, information technology systems and visual displays. They believe that Sweden has a well organised and knowledgeable health and safety organisation and that the legislation is of good quality. They find, however, that health and safety issues are not sufficiently integrated with other issues within enterprises and organisations.

The concerns during the 1970s were safety and work accidents. During the 1980s the issue of health and safety was enlarged to include occupational diseases, stress and psycho-social factors in the work environment. In the 1990s, knowledge about one's own workplace has been important as well as the consequences of changes in structure and policies that have to be dealt with. There is greater stress for those who are in work and there are changes in work roles with more individual responsibility and pressure for greater commitment and involvement. There is also a risk that changes in the labour market towards more flexible types of employment make it more difficult to work effectively with health and safety issues.

The unions would like a new agreement on occupational health services and are unhappy with the closure of the Working Life Fund. They hope that the grants for training and information in working environment matters for union members currently under discussion will not also disappear, and they hope that they will continue to particpate in the work on standardisation.

The future

The policy of the Government for the future is found in the remit of the National Board of Occupational Safety and Health for 1994. It stresses, amongst other things, the importance of dealing with issues at the workplace such as work organisation, the content of work, the physical and psychological stresses of work, and the work environment for women.

The goal for the future is to continue to work for the improvement of the work environment and to reduce the risks of ill-health and injuries throughout working life. There are great expectations of the Labour Inspectorate's new systems for inspection, the introduction of employers' internal control and the clear manifestation of the wish to make working conditions part of the production policy in many enterprises.

1 The context of national occupational health and safety policies in the United Kingdom

Economic structure

The United Kingdom has faced serious economic challenges during the period under review. Characteristic of Government economic strategies have been privatisation and denationalisation, reduced public spending and the market-testing of public authorities, encouragement for the self-employed and small and medium-sized enterprises, deregulation of the market and the legislative reform of trade unions. Employment has fallen during the period and there has been a continuation of the changes established in the previous decade in the distribution of employment between industrial sectors. Shifts in the pattern of work organisation from larger to smaller units and changes in human resource management have occurred. Decline in trade union membership and workplace representation continued, although no alternative form of employee organisation has emerged.

According to Department of Employment figures, in September 1993 there were 20.888 million people in employment. Approximately one quarter were in production and the other three quarters in services. Employment in agriculture accounted for only 281,000 people. The 1991 census of employment figures shows that nearly 45% of the people in employment worked in establishments of less than 50 workers, while 31% worked in establishments of 200 or more employees. The 1989 census of employment estimated that there were approximately 1.3 million establishments in Britain employing 21.7 million people. Of these, 1.1 million establishments were in the private sector, employing 15.9 million people, and the remainder employed 5.8 million people in the public sector.

The labour market

In January 1994 nearly 10% of the workforce, 2.79 million people, were unemployed. There has been a continual rising trend of unemployment since 1990. Unemployment is highest amongst the young and amongst ethnic minorities, with the Pakistani/Bangladeshi community, where a quarter of the economically active are unemployed, having the worst rate.

These statistics are a continuation of a number of trends seen over the past twenty years, during which time there has been a fall of nearly two million male workers in full-time employment and an increase of nearly 300,000 female workers. There has been a major shift away from employment in manufacturing to employment in services. The numbers of self-employed people have also increased dramatically, from nearly 1.8 million in 1986 to 3.2 million in 1992. Both male and female part-time work has increased during the last decade.

Industrial relations

Trade union membership has continued to decline since 1979. By 1992 it had fallen to 9.6 million, a fall of more than 25% from the peak 1979 level. It is the result entirely of a decline in male membership. In 1992, female trade union membership stood at 39%. At the end of 1992, the Trades Union Congress (TUC) - which is Britain's main trade union federation - had a membership of 7.3 million, again a decline over previous years and a reflection of the general decline in trade union membership. There are now 275 British trade unions. Nearly 80% of trade union members belong to the 20 largest unions. As a result of mergers between the larger trade unions, there are now a small number of trade unions with very large memberships. Well over half of the total membership of the TUC is concentrated in five trade unions.

Employers' organisations have also lost membership during the period. Only 1 in 8 workplaces recorded membership in a 1990 survey, compared with 1 in 4 in a survey in the same series carried out in 1980.

The most important changes in industrial relations in the 1980s and 1990s have been the decline in the representation of employees by trade unions and the decline in the coverage of collective bargaining, particularly in the public sector. These developments follow changes in the economy over the same period, especially the disappearance of many large manufacturing establishments. Other factors that have been important include the privatisation of much of the public sector, changes in internal business structures and developments in labour law. However, in workplaces where trade union representation and collective bargaining have persisted, little has altered. The changes that have occurred are a result of a decline in the proportion of workplaces operating the traditional system of industrial relations, rather than a universal decline across all sectors of employment. There is no indication that a new form of industrial relations has emerged to replace the traditional one.

Health and safety representatives and committees

The Safety Representatives and Safety Committees Regulations 1977 (SI 500) and the Safety Representatives and Safety Committees (Off-Shore) Regulations (SI 971) provide statutory rights for employee health and safety representatives and joint safety committees. The essential difference between the two sets of regulations is that SI 500 provides rights only for recognised trade unions to appoint health and safety representatives, whereas SI 971 entitles employees to elect health and safety representatives regardless of trade union membership. The main reason for this difference is the problem associated with trade union recognition off-shore.

Survey evidence of the implementation of SI 500 shows a growth in the use of joint safety committees during the early years of the 1980s and indicates that safety representatives were present in 35–40% of workplaces employing more that 25 workers. By the late 1980s, however, the spread of the implementation of the Regulations had stopped, with about 75% of employees covered by safety representatives. There is an association between the size of workplace and the presence of trade unions and safety representatives, and therefore a cause for concern that employees in smaller workplaces are without the benefit of representation.

2 Occupational health and safety policies and structures

Legislation

The Health and Safety at Work etc. Act (HSW etc. Act) 1974 provides the basis for the legislative regulation of health and safety in the United Kingdom in 1994. The Act enables the Secretary of State for Employment to approve Regulations in which the details of the specific legislative standards necessary to meet its requirements are spelt out. It also introduced the Approved Code of Practice (ACoP), an instrument which does not impose legal duties, but which sets out the means by which a legal duty may be accomplished. A feature of the HSW etc. Act 1974 is that it covers all people at work as well as risks to the general public arising from work. Furthermore, the Act and the Regulations are intended to be goal-setting rather than prescriptive legislation.

The Act established the Health and Safety Commission (HSC) as a tripartite national authority and the Health and Safety Executive (HSE) as the executive arm of the HSC, with the responsibility for achieving compliance with the provisions of the Act and those of the Regulations made under it. The HSE includes a number of previously separate Inspectorates whose inspectors have enforcement powers laid down in the Act. The HSE is a national body and, although some enforcement responsibilities are given to local authority environmental health departments (mostly for office and retail premises), the overall responsibility and authority rests with the HSE.

There are nearly 30 pieces of primary legislation (Acts of Parliament) that are relevant statutory provisions under the 1974 Act and more than 360 different sets of secondary legislation (mostly Regulations and Orders) that are also relevant statutory provisions under the Act. Risk assessment and control is the characteristic theme that binds all of the more important recent Regulations that have been introduced to implement European Directives, as well as some of the more significant Regulations that predate the recent spate of legislative activity stimulated by measures of EC origin.

Control and inspection

The HSE and the environmental health departments of local authorities are the main organisations responsible for the control and inspection of health and safety at work. The HSE enforces the 1974 Act and other health and safety legislation in over 650,000 workplaces, mainly in the industrial sector. It also covers temporary workplaces such as those in construction. The local authority environmental health departments enforce relevant health and safety legislation in around one and a quarter million other workplaces, mainly in the services sector.

The HSE employs 4,500 staff including over 1,500 inspectors and around 600 specialist and scientific staff. In addition to the general inspectors there are a smaller number of specialists who are recruited in the light of their particular knowledge and experience. Many of these specialists work in Field Consultant Groups and Field Scientific Support Units that are located in all of the seven regions into which the Field Operations Division is divided, and which provide specialist support to the general inspectors. The grant in aid to the HSC and the HSE in 1994/95 will be £190 million, and £192 million in the following two years. This represents a cut of £5 million in 1994 and £10 million in 1995/96 on previously announced levels.

There is an average of one factory inspector to 1,000 worksites for which the Factory Inspectorate is responsible. The equivalent figure for 1980 was one inspector to 420 worksites. In 1991, about half of the worksites had not been inspected for three years and nearly 70,000 had not been inspected for 11 years.

The original division of enforcement responsibilities between the HSE and local authorities under the 1974 Act has undergone some realignment under the Health and Safety (Enforcing Authority) Regulations 1989. This involved the transfer of a number of activities to local authority control which resulted in an increased inspection role for local authorities. There are about 5,900 local authority inspectors. It is estimated that the number of qualified staff undertaking health and safety at work duties is actually the equivalent of 1,440 people. Each local authority health and safety inspector has responsibility for about 3,000 premises. While there is a trend of increasing numbers of premises coming under local authority inspection responsibility during the period covered by this review, at the same time the numbers of staff working exclusively on health and safety during this period has fallen.

Local authorities on average inspect premises twice as often as the HSE – once every 2.5 years compared with once every 5 years by the HSE. They also on average issue 50% more improvement notices per inspector than HSE inspectors. This is despite the fact that, in general, premises inspected by local authorities are lower risk than those inspected by HSE inspectors.

On average there are some 3,000 prosecutions annually and some 35,000 to 45,000 enforcement notices issued by both the HSE and the local authorities together, compared with the 650,000 visits to workplaces undertaken by HSE and local authority inspectors in the course of a year. From its enforcement statistics, the HSE has recently estimated that:

- the annual average probability of a firm receiving an enforcement notice is 1 in 80
- the annual average probability of a firm being prosecuted is 1 in 800.

It has compared these probabilities with the annual probability of firms facing a reportable injury of 1 in 4.5; of facing a major injury incident of 1 in 27; and of facing a fatal injury incident of 1 in 800.

Occupational health services

Preventive services in occupational health and safety are characterised by a voluntary approach. An employer has considerable discretion with regard to the form of preventive service and the qualifications of his personnel. Preventive services are usually separate occupational health services and health and safety advisers and safety managers. A recent survey indicates that occupational health services that use professionally qualified staff exist in 68% of firms with more than 25 employees, compared with only 5% of firms with less than 25 employees. There is also variation according to industrial sector, with the highest percentage being found in manufacturing establishments. The use of health professionals is higher in the public sector. The proportion of employees and establishments covered by other health and safety professionals is not known, but it is probable that the pattern of their distribution would be similar to that of occupational health professionals.

The Employment Medical Advisory Service (EMAS), which is part of the HSE, is staffed by doctors and nurses with specialist qualifications in occupational health. It is responsible for all workplaces registered with the HSE. It consists of just over 40 doctors and 60 nurses, one doctor to every 20,000 workplaces and one nurse to every 330,000 employees. It provides specialist advice on occupational health, and advises on work and the training for work of people with disabilities. EMAS doctors and nurses investigate confirmed or suspected instances of occupational disease. They have the same inspection powers as HSE inspectors. There are statutory provisions that require medical surveillance for some people working with such materials as lead and asbestos or who are exposed to radiation. Statutory medical examinations under these provisions are carried out by EMAS doctors or appointed doctors approved by EMAS.

Information and assistance

The HSW etc. Act 1974 requires employers to provide information, supervision and training to employees and there are further requirements with regard to the rights of safety representatives to training and information in the Safety Representatives and Safety Committees Regulations (SRSC) 1977. The HSC has a statutory requirement to provide information and advice on health and safety. Most of the major sets of regulations have been introduced in recent years to implement the provisions of European Directives, including measures on the provision of information to workers and/or their representatives.

The major source of information on health and safety in Britain is the HSE. It estimates that it distributes eight million copies of guidance and other publications and answers 750,000 inquiries each year. In 1993 the distribution of information was one of the HSE activities that became subject to market testing, and retail distribution arrangements for priced publications were transferred from Her Majesty's Stationery Office (HMSO) to a limited number of private booksellers and distributors. Trade unions and the TUC provide another major source of information, and employers' organisations and the CBI are also active in producing information.

To support the requirements of the HSW Act 1974 on training, the additional measures of the Management of Health and Safety at Work Regulations 1992, which implement the provisions of the Framework Directive (89/391/EEC), make employers' duties with regard to training more explicit. In its policy statement on training, the HSC has also identified a number of key areas for development.

There is no national overview of the quantity of training undertaken in health and safety. Nevertheless, many organisations - including trade unions - are involved in the production and delivery of training and education at many different levels. The TUC has published annual figures of representatives attending its approved courses since the mid-1970s. In recent years the numbers have declined, reflecting the more difficult climate in which trade unions find themselves. There are no comparable figures for the extent of training provision for managers, but the HSE is currently undertaking a review of the total availability of this kind of training. In all of the professions involved in occupational health and safety, including occupational medicine, occupational health nursing, occupational hygiene, occupational safety, radiation protection and enforcement, professional recognition requires approved education and training to an advanced level as well as specialist experience.

The curriculum in primary and secondary education has undergone a major reorganisation with the introduction of the National Curriculum bringing various areas of the curriculum under statutory order. Several of these areas have recently benefited from HSE-supported curriculum development; for example, electrical safety within the statutory order areas of science and technology.

A Government initiative that is predicted to have a major impact on the training and competence of specialists in health and safety is the new system of vocational qualifications called National Vocational Qualifications (NVQs) and Scottish Vocational Qualifications (SVQs). These were introduced in the mid-1980s to help increase the competitiveness of

British industry. This initiative is intended to create a comprehensive system of vocational qualifications based on standards of competence agreed by appropriate employment sectors. The first of these competence-based qualifications relating to general safety is anticipated to be available from February 1995.

Research

According to its Annual Report, every year the HSE spends about 46 million ECU in support of its science and technology objectives. In addition to research commissioned and funded by the HSC, funding for research in occupational health and safety is available from other public bodies. Research in occupational medicine and related fields is funded by the Medical Research Council, in safety engineering it is funded by the Science and Engineering Research Council, and in social, economic, psychological and managerial aspects it is funded by the Economic and Social Science Research Council. Most of this work is carried out in universities and research organisations.

Monitoring

Employers have legal obligations to provide information on occupational injuries and ill-health under the requirements of the Reporting of Injuries, Diseases and Dangerous Occurrences Regulations 1985 (RIDDOR). It appears from the results of the Labour Force Survey (LFS) that less than one third of reportable non-fatal injuries at work are being reported under RIDDOR. Several other sources are routinely used for the national monitoring of occupational health, as well as RIDDOR data. These include prescribed diseases compensated under the Industrial Injuries Scheme, death certificates mentioning either asbestosis or mesothelioma, blood lead records from statutory medical surveillance and the LFS from 1993. New requirements for reporting accidents have recently been announced in a consultative document issued by the HSE, which anticipate European Commission requirements.

Economic incentives

The HSE has recently estimated that the overall cost to British employers of work-related ill-health and accidents is between 7.4 billion and 10.9 billion ECU a year. This is equivalent to between 5% and 10% of all UK industrial companies' gross trading profits for 1990, or between 205 and 435 ECU for each person employed. The cost to the economy as a whole is put at between 13 billion and 19 billion ECU, or 2% to 3% of GDP.

There has been a steady rise in the level of fines during the period of the review, which itself is a continuation of the rise during the 1980s. The most recently available data, those for 1992/93, show a 17% rise. This is a reflection of the introduction in the lower courts in March 1992 of an exemplary maximum fine of 24,160 ECU for serious breaches of the 1974 Act and failure to comply with an enforcement notice or court order, and the increase in November 1992 from 2,416 to 6,040 ECU of the maximum fine for other health and safety offences.

Occupational health insurance

Employers' and public liability insurance are statutory obligations on firms in Britain. Claims for compensation for industrial injury and ill-health rose threefold between 1983 and 1988. According to the Loss Prevention Council, a total of 704.3 million ECU was paid out by UK insurers (excluding Lloyds) in employers' liability claims in 1991. Payouts have risen from 375.7 million ECU in 1985 to 639 million ECU in 1990.

3 Health and safety outputs

Occupational accidents

Deaths from accidents at work have reached their lowest level since recording began. This is largely explained by changes in British industry – the shift from the high-risk industries such as manufacturing and mining towards service industries, although in a press conference on April 1st 1995 – to mark the 20th birthday of the HSE – the Chairman of the HSC claimed that 50% of the reduction in fatalities at work could be attributed to enforcement initiatives. The non-fatal injury rate has remained relatively stable since the present system of reporting began in 1986/87, with only a slight declining trend which is wholly attributable to changes in patterns of employment.

Occupational disease

With regard to occupational disease, the LFS data show that:

- 750,000 workers took 13 million days off work in 1989–90 because of what they regarded as work-related illnesses
- 730,000 in work were affected by work-related illness although they took no time off
- a further 820,000 retired and unemployed people reported that they were affected by the longer-term consequences of occupationally related illness.

General musculoskeletal disorders (discounting specifically upper-limb disorders) account for the highest number of cases caused by work (550,000), followed by hearing loss and stress/depression (110,000 cases each) and lower respiratory disease (nearly 90,000). The illnesses with the highest proportion of cases felt by respondents to be caused by work are pneumoconiosis (92%), hearing loss (85%), upper-limb disorders (69%) and skin diseases (65%).

A more recent reworking of the same data by the HSE indicates that at the time of the LFS nearly 6% of adults believed they suffered from an occupationally related disease. Higher risks of disease - such as musculoskeletal conditions, lower respiratory diseases, deafness and the long-term effects of injury - are found in manual occupations, with coal mining featuring as the occupation with the highest risk of disease. The construction industry, with a two-fold raised risk for all ill-health 'caused' by work, was noted for back problems and was responsible for high rates of skin disease in men. Work in metal and electrical processing, also nearly double the risk for work-related ill-health, is associated with high rates of deafness and vibration white finger. Asthma and eye problems are found with work in other materials' processing, and repetitive strain injury occurs where workers are engaged in repetitive assembly, inspection and packing.

Absenteeism

European comparisons using harmonised labour force surveys have shown that the United Kingdom had the second highest rate of employee absence in the European Union in 1990 when the most recent comparison was undertaken. The CBI estimates the direct costs of employee absence at 15.7 billion ECU a year. Another estimate has suggested that absence costs employers 7.2 billion ECU a year. The Industrial Society, as a result of its survey published in 1993, estimated that 200 million working days and over 10.9 billion ECU are lost every year due to absence. There is a general recognition that absence is under-reported by employers in all of the main surveys.

4 Assessment of occupational health and safety policies

Introduction

What are the significant developments and influences on occupational health and safety in the 1990s? Most of the period has been one of economic downturn. Government has been committed to an economic and political policy that is characterised by increased curbs on public spending, tightening of legislative control over the activities of trade unions, and privatisation of nationalised industries and public utilities, while at the same time promoting deregulatory measures to encourage enterprise and create a more flexible free market. Some of these features, such as the cuts in public spending, have impacted directly upon the national institutions and processes of occupational health and safety; others, the new legal challenges facing trade union organisation, for example, impact upon the situation in which health and safety regulation operates and hamper or enhance the relative roles of key players. Superimposed upon this national economic and political situation has been membership of the EU to which British legislators and regulators have had to become increasingly responsive. It is generally accepted by the social partners and the public authorities that the EU has been the major driving force for change in British legislation during the past five years.

The following examples reflect some of the main concerns that have been in the public domain in health and safety during the period of the review and which are prevalent at the present time.

Deregulation and occupational health and safety

In 1993 the Minister for Employment asked the HSC to conduct a review of health and safety legislation and advise the Government whether it was still relevant, whether it all remained necessary and whether it was possible to reduce the administrative burdens that it created for business, especially small businesses. The final report of the HSC Review recommends the removal of 40% of the volume of health and safety legislation. Seven pieces of primary legislation will be repealed and almost 100 sets of Regulations revoked. Key instruments to be repealed include the remaining parts of the two Acts that regulated health and safety at work before the introduction of the 1974 Act: the Factories Act 1961 and the Offices, Shops and Railway Premises Act 1963. However, the removal of all this legislation is recommended because it has outlived its purpose; it is a continuation of the process begun more than 20 years ago, and it does not represent a challenge to the main framework of legislative provisions. The Review refrains from recommending the repeal or revocation of legislation introduced to implement EC Directives; nor does it recommend the revocation of any of the significant regulations introduced since the 1974 Act. Furthermore, the Review found almost no support for exempting small employers or the self-employed from health and safety law. The Government has accepted the recommendations of the Review and has committed itself to putting them into effect.

Cuts in public expenditure and health and safety

While the deregulation debate has occupied a high profile, the regulatory system has also been restricted by the implementation of a series of cuts in public expenditure that have occurred during the review period.

Several aspects of public expenditure and regulatory policy need consideration together to appreciate their overall impact. In 1979 the Government instructed the HSC to account for the economic implications of proposals for new regulations. From that time HSE inspectors have been increasingly required to consider the cost benefit implications of their enforcement actions. In 1989 the Employment Secretary announced his intention to review the composition of the HSC, the result of which was a reduced role for TUC nominees. During the period 1975/76 to 1990/91, the proportion of HSE expenditure that was provided directly by the Government as grant in aid fell from 98% to 76%. Between 1987 and 1990 the percentage of the Department of Employment Group budget allocated to the HSE fell by 3.1%; further expenditure cuts for the HSE were announced in 1994, resulting in a £15 million reduction in the HSE budget between 1994 and 1996, and a consequent loss of 230 staff. An additional problem for the HSE is that, as a result of previous recruitment restrictions, the present and immediate future profile of the HSE is one where a high proportion of its staff are inexperienced, thus further reducing its ability to utilise staff to their full potential. Market-testing of the various branches of the HSE is also currently in progress.

These examples illustrate that changes in policy emphasising economic considerations in regulation-setting and enforcement practice have occurred at the same time as the role of the trade unions has been diminished. The period has been characterised by the increasingly stringent financial control of the HSE as well as privatisation of its peripheral activities and market-testing of its central activities, against a background of cutbacks in public expenditure during economic recession in related areas such as research and scientific support. It is apparent that changes in the post-1974 Act system for dealing with health and safety regulation are greater as a result of these developments than from the consequences of the Government's explicit deregulation initiative.

Penalties and enforcement

An important aspect of the development of preventive strategies in occupational health and safety is the question of the achievement of compliance with national and European regulations. During the review period there has been a rising trend in the level of fines for health and safety offences. There have been some other indications during the last few years that the courts are beginning to respond to health and safety offences more seriously than previously.

There are essentially three ways of increasing the effectiveness of external enforcement. One is to increase the resources of the inspectorates, the second is to change the enforcement policy and the third is to raise the penalties for breach of the law. It will be apparent that the first option has not been pursued in Britain, the resources of the HSE have been decreased rather than increased during recent years. The second option of changing enforcement practice has been the subject of some interesting discussion in recent literature. The HSE and local authority inspectors only initiate enforcement actions in a minority of cases in comparison with the number of inspections they undertake. The HSE argues that enforcement action is only one way in which it can achieve compliance and it is often not the most cost-effective method at its disposal. Its selective use is deliberate and deeply rooted in the evolution of the policies of the Factory Inspectorate, a fact that is borne out by the numerous critical studies of the development of health and safety enforcement practice in Britain. Its critics are mainly concerned with what they see as its failure to adopt a strict enough approach, but there is little hard evidence that a more adversarial approach would be any more effective than its present co-operative style, given the limited and reducing nature of the resources it has available. Comparative studies suggest that the British approach is 'at least as effective, if not more so', that the more adversarial American approach to enforcement.

The third option of increasing the penalties for offenders seems to be the way the legislature and the courts find easiest to respond to the contradictions of the deregulatory pressures of the free market and the requirement for society to protect employees and the public from the consequences of pursuit of profit at the expense of safety. The late 1980s witnessed a series of major disasters - Zeebrugge, Kings Cross, Clapham, Piper Alpha - amongst the most notable. Most of the inquiries into these disasters have pointed to failures of safety management. As a result, the courts appear to be taking a sterner view of serious health and safety offences. Some exemplary fines have been levelled, legislation has been passed increasing the general level of fines and there has been renewed interest in the use of manslaughter charges and the role of the police in the investigation of workplace deaths. However, the link between the level of fines and their deterrent value is not established. Furthermore, it is not necessarily the highest fines that are the least affordable.

A more specific aspect of enforcement practice that has attracted considerable attention recently is the nature of the enforcement action following a fatal accident at the workplace. There are three prosecution options available. The company may be prosecuted for a health and safety offence. Secondly, a company officer may be found guilty of an offence under health and safety law if the offence alleged against the company is considered to have been committed with his consent, connivance, or to have been attributable to any neglect on his part. Thirdly, prior to May 1993 the case may have been referred to the police or the Criminal Prosecution Service (CPS) for prosecution for the crime of manslaughter, if there was *prima facie* evidence of recklessness on the part of the company officer. For a successful prosecution to occur it was necessary to prove beyond reasonable doubt that recklessness on the part of the company officer was the cause of death. In a Court of Appeal ruling in May 1993 the law of manslaughter was redefined so that the test was not recklessness but 'gross negligence'.

Very few cases are referred by the HSE to the CPS, although it claims to be referring more cases than previously. The only successful prosecution that has occurred was brought as a result of police involvement at the scene of the accident.

That there are problems with the law of manslaughter when it is applied to health and safety cases is also evident from the failure of the prosecution of P&O Ferries Ltd as a result of the *Herald of Free Enterprise* Zeebrugge disaster in 1991. The prosecution of P&O failed because the obscurities of reckless or gross negligence manslaughter were compounded by the obscurities in the law of corporate criminal liability. One consequence of this failure has been the Law Commission's recent proposals on facilitating the prosecution of companies and individuals for manslaughter. The Commission has placed emphasis on the role of the company in being aware of risks and the existence within the company of systems for transmitting knowledge of risks and doing something about them. There is a close connection between the Commission's focus on the concept of risk and its central role in the new regulatory approach to health and safety following the implementation of the Framework Directive.

Trade union and employer involvement in health and safety in Britain

Employers' organisations and trade unions continue to play an active part in the tripartite structures created under the aegis of the HSC by the 1974 Act. Although there has been a slight reduction in the presence of TUC nominees on these committees and some examples of strong opposition by TUC nominees to measures put forward by the HSC being unsuccessful, there is little sign of any radical change in the overall system of tripartite representation. The CBI has sought to interest its membership in the positive aspects of health and safety at work through its efforts to promote safety management as an integral part of quality management. It has also highlighted the economic benefits of a healthy workforce through the development of health promotion programmes.

Changes in trade union membership and trade union organisation have caused the TUC to undertake a major restructuring, resulting in a much streamlined organisation. There has been some overall reduction in the resources available to health and safety in the reorganisation, but on the whole it has fared comparatively well and features as one of the priority areas for the recently relaunched campaigning TUC. The TUC has also identified some Common Action Priorities in health and safety on which it will lead new initiatives in collaboration with its affiliates, in the hope that the strong identification of health and safety with trade unions will help to boost their image.

The position that trade union workplace health and safety representatives currently occupy under British legislation is unique in Europe. It originates in the 1974–78 Labour Government's effort to legitimise trade unions as a means of employee representation in health and safety within the framework of collective bargaining. These measures were based on assumptions of trade union power that may have been valid in the 1970s, but are highly questionable in the 1990s. Surveys of the extent of safety representation in Britain show that by the late 1980s its spread had stopped and it was in decline in smaller and medium-sized categories of establishment. However, evidence from case studies also suggests that safety representatives are most likely to be effective in workplaces with strong trade union organisation.

The point about the research findings is that they indicate that in order to be effective, worker representatives need support. The institutions that provide this support at the workplace are trade unions. Research on joint regulation in non-unionised workplaces suggests that broadly the same conditions for effectiveness apply to non-union representatives as trade union representatives, the key condition being that of support. The problem for non-union representatives is where is this support to come from? Although the presence of trade unions has declined in British workplaces during the 1980s and 1990s, no alternative to their role at the workplace has emerged.

The British onshore legislative provisions on trade union appointment of safety representatives are out of step with the requirements of the Framework Directive which has no mention of trade unions in its provisions on consultation and information rights for workers and their representatives. In October 1995 the HSC issued a Consultative Document containing proposals for new Regulation to bring Britain into line with the rest of the EU on this issue. It appears that the proposed new Regulation will 'top up' the existing onshore provisions rather than replace them with a new system. This is a welcome development for the trade unions, and the extension of legislative rights to workers and their representatives might actually help British trade unions improve their representation in workplace health and safety, since it will help to create a new framework for representative rights that employers are legally bound to recognise.

Conclusions

During the last five years, prevention policies in occupational health and safety have faced considerable challenges from a variety of sources including the economic situation, new trends in business and work organisation, Government initiatives to encourage the free market, cuts in public expenditure and shifts in the balance of power between the main actors involved. At the same time there has been an unprecedented increase in the amount of legislation introduced, virtually all of it in response to the requirements of European Directives.

Although the general trend is one of reduction, the amount of human suffering as a result of workplace injuries and work-related diseases remains unacceptably high. Important new information on the extent of the problem of injuries and ill-health has become available, indicating the enormous extent of the under-reporting of injury and disease in the existing monitoring systems. The new figures have also enabled quantification of the economic costs of occupational accidents and ill-health in a more comprehensive way than previously attempted, and have highlighted costs that are much higher than previous estimates. The role of cost-benefit analysis and cost effectiveness generally has taken increasingly significant prominence in health and safety. Emphasis on cost effectiveness has been helpful in raising the profile of the economic consequences of failure to prevent accidents and ill-health. When coupled with the observation made by the HSE on many occasions - that the majority of preventable accidents could be prevented by better management - financial losses acquire even greater significance.

The courts appear to be taking a sterner view of serious health and safety offences. Some exemplary fines have been levelled, legislation has been passed increasing the general level of fines, and there has been renewed interest in the use of manslaughter charges and the role of the police and the CPS in the investigation of workplace deaths.

The British regulatory system in health and safety is the oldest in the world. Its most recent major overhaul and new strategy was provided by the HSW etc. Act 1974. In an effort to move away from the heavily prescriptive and externally enforced regulation-bound approach that had characterised British protective legislation since the nineteenth century, the Act created a system for the development of goal-setting regulation through the participation of representatives of the state, employers and trade unions. It also provided for the involvement of trade union representatives at workplace level. Arguably, it is out of step with the emphasis on individual rather than collective rights found in European legislation. The spate of recent European legislation has provided an opportunity for Britain to introduce another fundamental reform of its health and safety legislation by introducing a new statute. This opportunity has been eschewed. However, it has been publicly recognised that a major recasting of British provisions may become an eventual inevitability.

In all of the legislative activity that has been evident in recent years there has been considerable criticism that the measures introduced in Britain fall short of the requirements of the Directives they sought to implement. At the same time there is also a perception in Britain, articulated mainly by the HSE and by employers' organisations, that European legislation is actually more prescriptive than British strategy would wish, making it difficult to implement its requirements and at the same time retain the goal-setting character of the British approach. The reason given for this difference is the greater freedom of the courts in other Member States to interpret legislative provisions. Despite these concerns, the recent spate of regulations that have resulted from the implementation of the Framework Directive and its daughter Directives have adopted a consistent strategy based around the notions of risk assessment and risk management which themselves are goal-setting concepts and not prescriptive measures.

Both the social partners and the regulatory authorities agree that, with regard to the EU, the most likely development in the near future will be a period of consolidation during which time British industry and the enforcement authorities can get to grips with the consequences of recent legislative activity. While this is welcomed by the authorities and the employers, trade unions are less than sanguine about it. For them such a period poses important problems, not least because they believe that the major stimulus and support for the kind of social legislation in health and safety that they themselves want has come from the EC and not from within Britain.

European Foundation for the Improvement of Living and Working Conditions

The Identification and Assessment of Occupational Health and Safety Strategies in Europe
Volume 1: The National Situations

Luxembourg: Office for Official Publications of the European Communities

1996 – 220 pp. – 21 x 29.7 cm

ISBN 92-827-6641-1

Price (excluding VAT) in Luxembourg: ECU 16.50